THE RHYTHMS OF SOCIETY

Reports of the Institute of Community Studies

THE RHYTHMS OF SOCIETY

Edited by
Michael Young and
Tom Schuller

ROUTLEDGE
London and New York

First published in 1988 by
Routledge
11 New Fetter Lane, London EC4P 4EE

Published in the USA by
Routledge
in association with Routledge, Chapman and Hall, Inc.
29 West 35th Street, New York NY 10001

Set in 10½/12 pt. Sabon by
Mayhew Typesetting
and printed in Great Britain by
Richard Clay Ltd
Bungay, Suffolk

Library of Congress Cataloging in Publication Data

The rhythms of society/edited by Michael Young and Tom Schuller.
p. cm. — (Reports of the Institute of Community Studies)
Includes index.
1. Time — Social aspects. 2. Time perception.
3. Time management.
I. Young, Michael Dunlop. 1915– . II. Schuller, Tom.
III. Series.
HM73.R48 1988
306 — dc19 88-11444
ISBN 0-415-025338

CONTENTS

NOTES ON CONTRIBUTORS

Barbara Adam obtained her PhD thesis in February 1988. This is a study on time and social theory, exploring Anglo-Saxon and West German perspectives on social time and contemporary scientific approaches to natural time. Her research interests also include time as a source for post-Newtonian and Cartesian frameworks of meaning, and the role of time theory for feminist social theory. She has written papers on time and consciousness, time and feminist social theory, and time and health.

Johnston Birchall worked as a community worker and housing association manager before completing a PhD on housing cooperatives at York University. He was Research Fellow at the Institute of Community Studies in Bethnal Green, 1985–87 and is currently Senior Lecturer in Housing Studies at South Bank Polytechnic. He is the author of *Save Our Shop* (Holyoake Press, 1987) and *Building Communities: the Co-operative Way* (RKP, 1988).

Ronald Frankenberg is Emeritus Professor of Sociology and Social Anthropology, Professorial Research Fellow and Director, Centre for Medical Social Anthropology at the University of Keele. His most recent publication is on sickness as cultural performance and his theoretical interests include the refinement of concepts of illness, disease and sickness in terms of time and narrative and the development of a sociological/

anthropological critique of health behaviour.

Peter Hall holds joint appointments as Professor of Geography at the University of Reading and Professor of City and Regional Planning at the University of California, Berkeley. His main current research interest is the genesis and growth of high-technology industry world-wide.

Marie Jahoda is Professor Emeritus in Social Psychology at the University of Sussex. Her research has spanned the two periods of great unemployment this century. She is the author of numerous books and articles, most notably *Marienthal: The Sociography of an Unemployed Community* (1933, with P. Lazarsfeld and H. Zeisel).

Peter Laslett is Director of the Rank Xerox Unit on Ageing at the Cambridge Group for the History of Population and Social Structure and a Fellow of Trinity College. He is the author of, amongst other works, *The World We Have Lost* (1965, fifteen impressions), *Household and Family in Past Time* (1972), and *Britain! Be Your Age!* (forthcoming).

Tom Schuller is Senior Lecturer in Continuing Education at the University of Warwick. He has worked at OECD in Paris, the University of Glasgow and the Institute of Community Studies in London. He is the author of *Democracy at Work* (Oxford University Press, 1985) and *Age, Capital and Democracy* (Gower, 1986).

Ken Starkey is Lecturer in Motivation and Communication at the University of St Andrews. He was previously Research Fellow at the ESRC Work Organisation Research Centre, Aston University. Besides time, his main research interest is in the management of organisational change.

Nigel Thrift is Director of the Centre for the Study of Britain and the World Economy at the University of Bristol. His books include *Making Sense of Time* (edited, with T. Carlstein & D. Parkes, E. Arnold, 1978) and *Times, Spaces & Places* (with D. Parkes, J. Wiley, 1980). His chief interests are in multinational corporations, the City of London, the socialist Third World and social theory and time.

Michael Young is a sociologist who is known for his Bethnal Green Studies. He is the originator of the Consumers' Association and the Open University. He is a member of the House of Lords as Lord Young of Dartington. His most recent book is *The Metronomic Society: natural rhythms and human timetables* (Harvard University Press, 1988).

CHAPTER 1

INTRODUCTION: TOWARDS CHRONOSOCIOLOGY

Michael Young and Tom Schuller

Modern society is continually consulting its watch, so much so that it can fairly claim to be the most time-obsessed society there has ever been. Before electricity fundamentally altered the relationship between people and nature by taming night with light other societies responded to other beats. People who slept out in the open, or in caves and huts lit if at all by a guttering flame, were familiar with the sun, the moon and the stars in a way that is now quite uncommon. Gazing up at these other more mysterious lights everyone had to be an astronomer. The reappearances of the constellations populated by their different gods were signals which had to be attended to, as did the re-turnings of the year, the month and the day. People who still had night to contend with and learn from could hardly fail to see themselves as under the sway of the heavenly bodies. The Maya considered themselves as ruled by a series of gods which marched around and around in fixed cycles, with different but very precise periodicities; between them the recursive gods ordained what was proper in any day, in any year or in any more protracted cycle.

But for the Maya, and indeed for most of our ancestors before the invention of the mechanical clock, and especially before 1820 when Pall Mall became the first street in the world to be lit by gas, the intervals that mattered were the intervals established by the earth and the moon. The beats were comparatively slow. It has only been as the tick-tocks have been made to sound in every cranny of every institution that the critical intervals have become so short. The first clocks, in the thirteenth century, had only hour hands and even their invention made a large difference; the nights could from then on be thought of as being divided into hours of the same length as the days and the duration of unconsciousness measured as precisely as the duration of consciousness. Mankind was already launched on its way to a miniaturising standardisation which has since swept the world, the hour hand being later augmented by a minute hand, and the minute hand by a second hand. Trains leave (or advertise themselves as leaving) at 14.17; television programmes boast of their fidelity to punctuality by displaying second hands which tick around to the minute's midnight when the soap opera has to start up all over again; athletes are timed to a hundredth of a second, and in laboratories the scientists peering down their microscopes subdivide their seconds much more finely than that. It is a great game we are all playing, and more than a game. We have become like Alice's White Rabbit, always aware not of what time is but of what time it is, always either late or afraid we are going to be. Whether late or not, we are chronically short of time, and conduct our breathless lives accordingly and we become all the more pressed, the more we take away from nature some of its power over the timings of our environment and arrogate the power to ourselves. We hardly think twice about forcing our own new time upon other creatures, as, for instance, when we plan to use melatonin to require lambs to be born throughout the year instead of their being confined to the spring. We treat ourselves or are treated by others in the same way, as when doctors cut down on the number of babies born on Saturdays and Sundays because of the inconvenience of it to themselves. The electricity which has followed gas has also made it possible to move across the natural frontiers and 'colonise' the night (as Murray Melbin has put it, 1987) by forcing it to submit ever more fully to the government of the same clock that rules the day.

If we are not only obsessed with clock-time but getting more so, it is all the more strange that social scientists should have done so little to make time one of their special apprehensions. This must be because their subjects – other people

– do not prompt them to do so. No one can escape from time – everything people are and do is bound up with it – but they are not continuously aware that this is so. *The* truism about time is that we take it for granted. We take our clocks for granted, hardly ever remembering what a struggle it was with the aid of large cardboard dials at home or in the schoolroom to gain clock-literacy by learning to 'tell' the time; we take for granted that we have to go regularly to school every day at 9 o'clock from the 'time' we are five years old; we still take for granted that we shall 'retire' at some point, and die at some point thereafter; we take for granted that governments last so many years and the next instalment of another serial so many minutes; we enjoy our meals no less and may even enjoy them more because, hungry or not, we are eating them when we have eaten them before. The everyday authority of time is, even in a permissive society, so complete that it rarely appears as problematic, and if it is not problematic to the people who are the subjects it will not be all that problematic to the other people who are their not so eager students.

If we are right about time as a taken-for-granted obsession that is all going to change, as what we have taken for granted no longer suffices, and in a small way perhaps this book is a sign of the times. So is the Association to which the authors belong, the Association for the Social Studies of Time. This is an inter-disciplinary body, as shown by the contributors themselves. Time is not a subject like other subjects. Being ubiquitous, it straddles all disciplines, not just in the social sciences but well beyond, and, insofar as it is recognised as a common interest, time is an actual and even more a potential unifier, in life as well as in death. It is difficult, when considering it, to be too ardent about defending and upholding the traditional boundaries between disciplines. It is understandable that Nigel Thrift and Peter Hall should be geographers, Peter Laslett an historian, Marie Jahoda a psychologist, Ronald Frankenberg a social anthropologist, Ken Starkey an industrial sociologist and Johnston Birchall, Barbara Adam and ourselves as the editors sociologists plain.

Michael Young and Tom Schuller

THE PUZZLE OF SOCIAL STRUCTURE

Whatever their discipline, if social scientists are interested in time they must be prepared to puzzle in new ways about the relationship of the present to the past. Except when under challenge, people readily talk about and even think about the present as though it were in a separate category (or *is* a separate category) from the past. Although we cannot say that this 'moment' or that is, definitely, the present moment for it has partially disappeared into the past before it can be identified – say now and it has already gone – we can with the aid of our memories imagine a present which is abstracted, however briefly, from the passage of time, or, however arbitrarily, measure off a portion of the continuous flow as being a minute or a year, even a millennium, and in a particular context seize on that as the present. We are, as well as have been. It sometimes seems, in other words, that what we are has a little measure of permanence to it – that what we are is what we were and what we in all probability will be for a little longer. If it is proper at all to freeze any part of the flow into a series of past presents to compare with *the* present, or with a series of present presents, then we can with the aid of our memories notice that there has been a little change in between. The tree, or the university, or we, are not exactly the same from one present to another but in some respects, perhaps in very major respects, we are the same. The tree, the university, us, have a sufficient degree of permanence to hold chaos at bay.

Along with other social scientists, sociologists have enjoyed their own specific kind of puzzlement, viewing from a kind of distance the manner in which they, and their subjects, consider the events in which they are involved as being either continuities or discontinuities. The continuous present (rather than saying that something happens, we can say it *is* happening in order to emphasise that every present is a becoming) has been given various descriptions. It has been called static to distinguish it from dynamic – the one word meaning that something has *not* changed over time, except time, the other that something, or a lot of somethings, have changed; or, less colloquially, the oppositions counterposed have been synchronic and diachronic. Whatever the pairs of terms, the notion has been that in a society or a personality the continuities are organised into a 'structure' which persists even while it changes or, perhaps, *because* it changes. We see no harm in making that distinction. There is no reason why schools of sociologists should not concentrate, if so minded, on the continuities in the structure, as in modern times the followers of Parsons have tended to do. It is only unfortunate if with that emphasis the discontinuities are neglected. Likewise, there is no reason why schools of sociologists should not concentrate on the discontinuities, as in the last century and this the followers of Marx have tended to do. It is only unfortunate in that case if (contrariwise) the continuities are put on one side as if of no account compared to the flux and the jumps in the flux from quantity into quality.

If the time-obsessions of modern society become reflected in sociology, and go a lot wider than a concern with the kind of measurement which clocks have introduced, our hope is that some new light will be thrown on old distinctions like that between the static and the dynamic, and some new understandings gained of what can be envisaged as the structure of society. Once time is recognised as a continuous flow – with the essential continuity being in the flow itself – what is being observed cannot be anything other than change, continuous change. This is not to say there is no pattern, no structure, in the welter. But whatever has pattern, structure, an appearance of the static, is made up of change wrapped within change. We think that most of the authors would accept that they are not only interested in time for its own mysterious sake but for the sake of the unification it may bring about between the static and dynamic, the persisting and the disappearing.

One consequence of the customary division between the two is that institutional boundaries have been built around it. As Giddens has said,

> On the basis of this division, sociologists have been content to leave the succession of events in time to the historians, some of whom as their part of the bargain have been prepared to relinquish the structural properties

of social systems to the sociologists. But this kind of separation has no natural justification with the recovery of temporality as integral to social theory: history and sociology become methodologically indistinguishable. (Giddens, 1979, pp. 7–8)

We would not go along with that word, recovery, which suggests that all we have to do is to recover an ideal past. Social theory was, in our view, never in that happy state, with time integral to it. But it might in the future be happier with the past which was the present and with the future which will be the present before it too becomes the past.

On one point we are fully in agreement with Giddens: history and sociology should be kept as much together as allowed by the institutional walls which have been built up in and out of the universities. Peter Laslett is one 'historian' who would agree. In his essay on 'Social Structural Time' he has declared his task as being one of 'historical sociology'. This is not the last word but almost the first word on comparative rates of change treated in this way. Many of the changes which concern Laslett are cyclical and, if so, they are at once in the past and the present. Cycles, imagined or real, have carried many historians into the present, sometimes with a bump, sometimes without. There is no bump for Laslett but there may be for some readers who are brought up short by his scheme. Very few will share the 'assumption that there is a single pace of change, the political pace of change'; but if they do not, and so accept that there are many different paces, they are up against the Laslett problem. Who does not accept the conclusion may be hard put to offer his or her own.

The other two essays in historical sociology are, as befits this book, by geographers. Peter Hall has written about cycles as well, in his essay about one of the slower cycles, with a long period to it. The more and more elaborate records that we delight ourselves with in industrial societies enable us to search through the smoke for the signs and portents that were once reserved for the necromancer. Modern necromancers must bring statistics to their support, and Hall has produced a summary of the debate which, since the eclipse of Keynes, has raged fiercely about the 'long waves' which he did not acknowledge but to which Schumpeter gave such prominence. Keynes's cycle of popularity is on a downswing and Schumpeter's on the up, carrying with it a new interest in the cycles featured by Schumpeter which were named after the Russian economist, Nikolai Kondratieff. The upswing in the attention given to this school is the counter to the downswing in the world economy in the 1970s and 1980s which does not lend itself well to a Keynesian explanation. But bring in the longer view – the 50 to 60-year swings of Kondratieff – and the 1970s and 1980s can be viewed as a re-run of the 1920s and 1930s. Kondratieff provided the figures, Schumpeter the explanation. Because of the accumulation of capital, the profit rate falls, as Marx proposed, and out of the resulting crisis capitalism is repeatedly rescued by the entrepreneurs who supply a new wave of innovations to drive (or re-drive) the economy. But in the course of responding, capitalism could turn itself into something else, with monolithic public and private corporations which are indistinguishable from each other all propelled forward by faceless managers responsible only to each other.

The next essay, by Nigel Thrift, is not about cycles but about another traditional interest, secular change. This interest is shared with natural scientists. The subject matter for physicists is as much the macro of the universe as the micro of the atom, and the universe cannot be questioned without questioning time. Where did it begin, and how, and how (since we are still there to put the questions) is it still developing? The light which brings information now about how things were millions of years ago provides the fullest of historical records that physicists and others cannot help puzzling over. Biologists have also had to do the same, examining the record of the past left behind in fossils, in creatures which were evidently our ancestors, and in the foetuses which recapitulate some of our evolutionary past. Historians and social scientists generally, since the eclipse of the Enlightenment, have not been as much concerned with evolution as physicists and biologists; but, even if no such clear pattern stands out as it seems to in nature, they have been almost as much intent on the course of change. So much is this so that this might be thought of as the traditional interest of the social sciences in time.

The tradition is well represented in Thrift's essay. As a geographer he has as it were crossed over from space to time and given us an account of time-consciousness in the middle ages, from 1100 to 1300. This period is not usually portrayed as having much affinity with what are now, until they are superseded, called 'modern' times. When comparisons are made the accent is usually on the differences rather than the similarities, the middle ages being a golden age when people, before the clock carried them into a new enslavement, could be at one with the rhythms of nature. Thrift's picture is different. He has shown that people were as adept then as they are now in imposing harsh time-disciplines upon each other, with the aid of nature's bench-marks, as always, but also with the powerful aid of the religious calendar and the bells which tolled from the towers of religion. When the curfew bell was rung, at 9 o'clock in large towns, all fires had to be doused and lights extinguished. The modern time-obsessionalism, if it clearly has features all of its own, is not wholly unique.

Then follow two essays on power, a matter which relates very clearly to the central interest in social structure which we mentioned earlier. Power is a so-far permanent feature of society, and continuities are its essence. Power can last a lifetime and, in the hands of an elite group of some kind, can last longer than that, and, as Ken Starkey points out, a crucial kind of power is power over time. His basic perspective is Marxist. According to Marx, capitalism depends upon the exact control of time, so that more can be wrung out of it by the lengthening of the working day or the intensification of work. 'Man is nothing; he is, at most, the carcase of time.' The thrust of the employer is clear enough – for more productivity and more profit. On the other hand the worker can either allow his work to be degraded by higher productivity and settle for more money in compensation, or put up a fight for more meaningful work. Starkey uses as his illustration the dispute between British Rail and the Engine Drivers (ASLEF) over 'flexible rostering' and the retention of the eight-hour day. At issue was not simply the number of hours to be worked, nor even the mechanics of their allocation and the level of their remuneration. Whilst the employers sought to reach an agreement based on a conception of working time as a unitary commodity amenable to straightforward division, the union struggled to maintain a broader perspective which brought two other elements into the reckoning: the interaction between working time and family or leisure time, and the relationship between the control of working schedules and the exercise of occupational skills. The case study explores conflict over the interpretation of the meaning of time at work, as well as the rewards to be paid for it. For Starkey the implication is that in the future more and more disputes will turn on the same issues.

For Ronald Frankenberg time is just as central in medicine as it is in industry, with the doctors and other medical staff being as much (or more?) the time-controllers than employers. When patients are taken to hospital they are accompanied by their bodies, but both then and afterwards their bodies (and even to some degree the pain in them) are to some extent taken away from them and transferred over to the power of the professionals. They have to submit to the routines of the institution, being woken too early for the patients, for example, although perhaps not early enough for the nurses. Above all, they have to wait until this or that correct moment is reached in a sequence prescribed by their doctors and their assistants. In or out of hospital, patients are forced to belong to a 'waiting culture', in which if the illness lessens it will have to do so by numbers, waiting for a GP appointment, waiting in the GP's surgery, waiting for an out-patient appointment, waiting in the out-patient department on the day, waiting for an admission, when admitted waiting for the treatment, and waiting for the discharge which signals a decisive step towards recovery. So much do people have to wait that the first question a patient asks is 'how long' – how long to feel better, how long to get in, how long to get out, before recovering the self-control that has become customary outside hospitals and surgeries and the plethora of 'waiting rooms'.

Those who are not called patient for nothing have to behave as though they have plenty of time to spare while, in line with the 'cultural performance of illness', the doctors behave as though they have little or none to spare. They are the opposite of the patients – not the inpatients but the impatients. They are not supposed to wait at all, but to exercise their ritualised power they may also have to sacrifice a good deal themselves. The bleeps that they carry around with them, in and out of hospital, mean that the public persona and the private person are never separated, from the moment when they are taking the 'history' of a new patient to the moment when they fall into their own beds at night, and perhaps not even then. The doctor has, on top of his or her skill, a symbolic value which is as significant as that of the priest, and the value is dependent upon the manner in which this particular duality of time – the patient's and the doctor's – is dually (if rather one-sidedly) manipulated. Frankenberg is an introduction to the kind of study of time which is being pursued by the medical anthropologists.

With Marie Jahoda, the perspective shifts more decisively to the individual, and to the structuring and re-structuring of time as an experience with which individual people are continuously engaged. This is most obvious when there is a change in a structure. In perhaps the most famous study of the matter carried out in this century, at Marienthal, an Austrian village where virtually everybody was unemployed, Jahoda and her colleagues (1933) showed that everything slowed down. When there was 'nothing' that had to be done, people's walking speed slowed down, they stood around, they were unpunctual even for the few occasions when timing still mattered. At least that is what happened to the men in the village. For the women it was different. Poverty imposed upon them an even greater burden of time-consuming drudgery. For both sexes, as Jahoda says, 'there was nothing to look forward to, no hope and no planning for the future'. This was only one of many studies by psychologists which have shown how time is patterned by individuals and by the groups to which they belong. Seen in one light, time is very much a social construct, formed by individuals who are also more than individuals.

The 'sense of time' is treated differently by different psychologists. Jahoda considers Piaget and Freud the two greatest psychologists of the century. Piaget shows how 'time' has to be learned as it is differentiated from space, while Freud has revealed a new dimension in which the past lives on – clings on – in the present. Since Freud the clingingness of childhood has been studied by many other psychologists, and yet, despite all that, Jahoda – herself one of the outstanding social psychologists of the century – ends by regretting the neglect of time in her discipline. To say that is to propose a future in which it receives more attention, and more systematic and more cumulative attention than it has had so far. This could be the theme, implicit if not explicit, of any of the essays in this volume.

The next contribution is different again. Johnston Birchall is concerned with the ethical questions raised by any consideration of fraternity. Fraternity is not a momentary sentiment, but a continuing one, and it is one of the bonds which underpin society. But the fellow-feeling which most people have for each other to some minimal extent, and a few people have to a very large extent, is interpenetrated by a competitiveness which pulls the other way. The ethical and sociological problem is about the balance between these opposing forces, and their reconciliation or non-reconciliation in the body politic. What Birchall is able to show is that the co-operation (in practice and in spirit) and the competition which is also ever-present are both the work of time. They are both social constructs, instilled in people more by their persistent customs and habits than by the play on their reason. The competition is conducted within a framework of co-operation which lays down the rules which the competitors have ordinarily to obey, and both are conducted within the more all-embracing framework of behaviour which is taken for granted. The widest possible co-operation, continuously operative in the present, is dependent upon what has been observed in the past. Fraternity in general, and in particular, is an unconscious affair. As Birchall sees it, 'the competitive individualist plays on the surface of things while the deep structure of co-operation provides the supporting framework'.

Finally, Barbara Adam presents a direct challenge to social scientists to review their approaches to time in the light of changes in the conceptual apparatus of natural science. Social scientists, she argues, where they have taken an interest in time at all have marked themselves off from natural scientists by deploying two oppositions or antitheses: between the qualitative and the quantitative, and the subjective and the objective. In each case they have seen themselves as concerned with both terms of the pair in contrast to natural scientists who draw only on the latter term in each case. Social scientists, moreover, argue that all time is social time, thereby denying any role to natural time. Such a position is too static; the dichotomies are based on an outdated model of the natural sciences and an inadequate understanding of natural time.

Adam substantiates her argument by exploring the shifts which have taken place since Newton in the understanding of natural time, and reproaches social scientists for their failure to grasp the dynamics of conceptual innovation. One of the most bewildering features of time is its ability to transform itself into so many different guises. Measurement is one of the tools with which mankind seeks to bring order to nature; time is both an indispensable instrument of measurement and a key element to be measured. It is both subject and object, so much so that it is almost swallowing its own tail. As we get used – in principle, at least – to the concept of relativity and to post-Einsteinian advances in natural science, we can see how time as a quantitative measure has emerged as a human creation, as a part of social life in a particular context. Conversely, qualitative time is not a peculiarly human invention or characteristic, but a fundamental aspect of nature. Even the phenomena associated with that 'hardest' of sciences, physics, have their own rhythms which cannot be reduced to a standardised beat. However, Adam does not suggest that these revelations lead to a simple reversal of concepts, a straight half-turn of the wheel, as it were. The point is rather that the multiplicity of times can only be grasped and understood by refusing to dichotomise 'nature' and 'society'; the way humans symbolise time and relate to it is a specific explicated form of something that is implicated in all nature.

Adam suggests that anyone dealing with time is obliged willy-nilly to become in some measure a philosopher. She turns to a professional philosopher, McTaggart, for his well-known distinction between two types of temporal relations: the B-series, referring to the sequence in which events occur, and the A-series which deals with the pastness, presentness and futurity of events, and is thus dependent on the temporal position of the observer. This distinction has been mulled over, contested and rephrased by several authors since. In line with her previous argument, Adam draws attention to how Prigogine has established the A-series as a law of nature, and uses this to emphasise the interdependence of the various domains.

This last essay brings together several of the topics that are treated in earlier chapters. The rejection of straight dichotomies; the need for an historical perspective to capture the changes which are taking place even in our views of change itself; the appeal across disciplinary boundaries; the inherent limitations of separating out distinct parts of a rhythmic whole – all are themes which run through the volume.

The title *The Rhythms of Society* embodies a hope, the hope that the social sciences will make a great deal more of these rhythms than they have so far, and to end the introduction we should say something further on this score. Our cue comes, as much as from anything, from chronobiology. The *International Journal of Chronobiology* defines its subject as 'objectively quantifying and investigating mechanisms of biological time structure, including rhythmic manifestations of life' (1973, p. 35). It deals with the multiple rhythms of biological organism: infradian, circadian, circalunar, circatidal, circannual and so forth. If by analogy there were a chronosociology, its concern would be with the rhythms of social institutions and processes.

But we are not implying that the observations of the biologists can be neatly transposed. Human beings may be like other creatures, but they are also profoundly unlike, above all in the quality of their subjective experience. Rhythm is not just ordained from outside the person, it comes from the way people order their experience. Seen in this light, rhythm is a matter of the interplay between repetition and surprise. Sixty years ago the literary critic I.A. Richards analysed it as follows:

> Rhythm, and its specialised form, metre, depend upon repetition and expectancy. Equally where what is expected recurs and where it fails, all rhythmical and metrical effects spring from anticipation. As a rule this anticipation is unconscious. . . The mind, after reading a line or two of verse . . . prepares itself ahead for any one of a number of possible sequences at the same time negatively incapacitating itself for others. The effect produced by what actually follows depends very closely upon this unconscious preparation and consists largely of the further twist which it gives to expectancy. *It is in terms of the variation in these twists that rhythm is to be described . . . . This texture of expectations, satisfactions, disappointments, surprisals, which the sequence of syllables brings about, is rhythm.* (Richards, 1925, p. 134, 137, our stress)

There is the same tension, between predictability and surprise, in the analysis of past and present events and processes. Repetition itself is not enough to supply rhythm, other than the most primitive kind. There must be variation too, and all social phenomena exhibit friction and conflict between the sustaining of regularity and the intrusion of innovation.

Our perception of rhythm also depends on the level of activity upon which we focus within social structures. Some rhythms – those described by Hall in this volume are a prime example, Braudel's *longues durées* another – require a very wide-angle lens to bring them into focus, and would be completely invisible to the social scientist peering at a more limited segment of historical time. Conversely, the broad sweep of history at the macro level cannot capture the infinity of pulses that sustain particular households, housegods, workplaces or individuals in their daily lives. As Martins puts it:

> As one switches from one date-class to another, let us say from more micro to more macro histories or historicities, one gains information – 'perspective' – but also loses information – events, properties and phenomena which cannot be analysed or explained in terms of another date-

> class level of study. It is logically impossible to unify and totalise any two contiguous histories pertaining to different date-classes (or if you like 'times'). (1974, p. 264).

He concludes that 'No one mode of historical time is privileged' (ibid., p. 265).

There is no hierarchy in the authoritarian sense, rather what Arthur Koestler has called a 'holarchy', or system of holons. A holon is a

> stable, integrated structure, equipped with self-regulatory devices and enjoying a considerable degree of *autonomy* of self-government. Cells, muscles, nerves, organs, all have their intrinsic rhythms and patterns of activity, often manifested spontaneously without external stimulation; they are subordinated as *parts* to the higher centres in the hierarchy, but at the same time function as quasi-autonomous *wholes*. They are Janus-faced. The face turned upward, toward the higher levels, is that of a dependent part; the face turned downward, toward its own constituents, is that of a whole of remarkable self-sufficiency. (Koestler, 1978, p. 27, stress in original).

In one sense, then, social units have their own rhythms and march to their own beat; from another point of view they exist as part of a larger score.

Moreover, the ability of a person to perceive rhythm changes over time, as experience is accumulated. A piece of music which to the inexperienced ear is a random sequence of notes may have a clear shape and significance for the cultivated listener. Individual producers and consumers generally still have little idea of how their labour and their purchasing habits contribute to a wider economic pattern which brings together offices and high streets and villages throughout the country and the world; yet people today have more grasp of economic interrelationships than did their forbears.

The complexity of the pattern grows continuously, for instance as a result of the internationalisation of the economy and the intensification of financial transactions on a round-the-clock basis. But there is no reason for assuming that our understanding of the rhythms of society is making steady progress. Rather than catching up, it may be losing ground; but it does not need to.

REFERENCES

Giddens, A. (1979), *Central Problems in Social Theory*, London, Macmillan.

International Journal of Chronobiology, 1973, vol. 1, no. 1.

Jahoda, M., Lazarsfeld, P.F. and Zeisal, H. (1933), *Die Arbeitlosen von Marienthal*, Leipzig, Hirzel.

Koestler, A. (1978), *Janus*, London, Hutchinson.

Martins, H. (1974), 'Time and theory in sociology', in John Rex (ed.), *Approaches to Sociology*, London, Routledge & Kegan Paul.

Melbin, M. (1987), *Night as Frontier – Colonizing the World After Dark*, New York, Macmillan and The Free Press.

Richards, I.A. (1925), *Principles of Literary Criticism*, London, Kegan Paul, Trench, Trubner.

CHAPTER 2

SOCIAL STRUCTURAL TIME: AN ATTEMPT AT CLASSIFYING TYPES OF SOCIAL CHANGE BY THEIR CHARACTERISTIC PACES

Peter Laslett

It is the historian who has to be concerned with change over time and in so doing to make evident that the proper description of the task to be undertaken is historical sociology. Most of the history which people remember is political or military, and all that the citizen usually gets to know about the past is in narrative form, concerned above everything else with reigns, ministries and wars; kings, statesmen and generals. There are traces, however, mostly left unconnected with politics and wars, of religious and economic facts and changes, and of developments in literature, in communications and in technology, even in demography, if practically never of developments in the family.

Once it is demanded, however, why political events or military events succeed each other as they do, at the speed with which they happen; or why they tend to repeat themselves; or how they are connected with non-political events and changes such as those we call economic or religious or aesthetic; then it becomes obvious that political change is a part, an aspect, a facet – the words never seem to catch the reality – of the large and largest whole which is the entire society in its temporal vicissitudes. Hence analysis of a kind which would help us to understand why politics change in the way in which they do, at the speed at which they do, has to be called historical sociology. What is true of political events and political change must be true of events and changes of other descriptions. Changes in such fundamental institutions

Table 2.1: List of descriptive modes

1 *Event-like*	
1a event dominated.	1b event expressive.
1c event linked in other ways.	1d independent of events.
2 *Elite-associated*	
2a confined to elite.	2b dominated by elite.
2c elite linked in other ways.	2d independent of elite, society inclusive.
3 *Violent*	
3a necessarily violent.	3b usually violent.
3c occasionally violent.	3d never violent.
4 *Repetitive*	
4a necessarily repetitive.	4b occasionally repetitive.
4c never repetitive, unique.	
5 *Cyclical*	
5a always cyclical.	5b occasionally cyclical.
5c never cyclical, always linear.	
6 *Reversible*	
6a always reversible.	6b usually reversible.
6c never reversible, final.	
7 *Pace variable*	
7a always liable to change in pace.	7b usually liable to change in pace.
7c not susceptible to change in pace, constant.	
8 *Transformative*	
8a never excludes previous state.	8b occasionally excludes previous state.
8c always excludes previous state, entirely transformative.	
9 *Causative across categories*	
9a never leads to change across categories, self-contained.	
9b occasionally leads to change across categories, potentially deep seated social change.	
9c always leads to change across categories, fundamental social change.	

as the family should never have been left out.

It is the thesis of this contribution to the study of the sociology of time that the primacy of politics in the study of change, change going on in the present as well as change which is now complete, has a particular effect on social perception and social analysis. It is responsible to a considerable degree for the assumption that there is a single pace of change in collective activities, the political pace of change. When made explicit, it is obvious that this is an error, an understandable error as we shall see but an error of order all the same. I intend here to try and set out the categories of change which may become conflated with the political category of change, or confused with it. The major task, then, is one of social description, description of the types of activity whose paces of change concern us.

A lengthy and tedious task this will have to be: anyone who has ever attempted to write out a set of social categories in full will agree with alacrity. The only expedient for abbreviating and undertaking, and so making it manageable for the analyst and comprehensible to the reader, is to construct a table, a matrix of rows and columns. My attempt to anatomise the paces at which social change goes forward in the various areas conventionally recognised as appropriately distinct for analytic purposes is set out in the double tabular spread which accompanies this piece of prose. 'Accompanies' is not quite the word, because the pattern of nominal references contained in the table, and above all the orders of pace listed on the left hand side, make up the body of my contribution to our subject. The rest of what I have to say is a commentary upon the four orders of pace, with their suggested sub-division, and the nine columns facing them devoted to modes of change.

This is the last in a series of attempts at an anatomy of a hierarchy of the varying paces of types of social change, and yet the outcome seems still to be almost as much a matter of scantily founded assertion and guess-work as it was in the beginning. Lack of evidence, or of evidence known to me, is one reason for its insufficiency. A second is the want of previous reflection upon the topic and of discussion of the principles related to it, which in turn may help to explain the

Table 2.2: Types of social change

	Pace order	*Assigned descriptive modes*								
		1	2	3	4	5	6	7	8	9
		Event-like	*Elite-associated*	*Violent*	*Repetitive*	*Cyclical*	*Reversible*	*Pace variable*	*Transformative*	*Causative across categories*
A.	**Change of fast-paced order**									
A (i)	*Political* change	1a 1b	2b	3c	4a	5b	6b	7c	8a	9b
A (ii)	Change in *fashion*, apparel, etc.	1a 1b	2a	3d	4a	5a	6a	7a	8a	9a
A (iii)	Change in *opinion*, political, social, intellectual	1a 1b	2b	3d	4b	5b	6a	7a	8a	9b
A (iv)	*Military* change, by organised force	1a 1b	2b	3b	4b	5a	?	7b	8b	9b
A (v)	*Insurrectionary* change	1a 1b	2c	3a	4b	5b	6a	7b	8a	9b
B.	**Change of medium-paced order**									
B (i)	*Economic* change	1c	2c	3d	4b	5b	6b	7d	8b	9b
B (ii)	Change in *communication*	1d	2c	3d	4c	5c	6c	7d	8b	9b
B (iii)	*Technical* change, in production, etc.	1d	2c	3d	4c	5c	6c	7a	8b	9b

B (iv)	*Demographic* change	1c	2d	3d	4b	5b	6a	7b	8b	9b
B (v)	*Intellectual* change	1b 1c	2a	3d	4c	5b	6b	7a	8b	9b
B (vi)	*Aesthetic* change	1b 1c	2a	3d	4c	5b	6b	7b	8a	9a
B (vii)	*Institutional* change	1c	2a	3d	4b	5b	6a	7b	8c	9b
C.	**Change of slow-paced order**									
C (i)	*Constitutional* change, in overall public organisation	1a	2b	3d	4c	5c	6a	?	8b	9b
C (ii)	Change in *belief*, other than religious	1c	2b	3c	4c	5c	?	7b	8b	9b
C (iii)	Change in *religion*	1b 1c	2c	3c	4c	5b	?	?	8b	9b
D.	**Change of very-slow-paced order**									
D (i)	*Normative* change	1d	2b	3d	4c	5c	6c	7c	8c	9c
D (ii)	Change in *relations of production*	1d	2b	3c	4c	5c	6c	7c	8c	9c
D (iii)	Change in *social structure* — micro social structure, family, etc.	1d	2d	3d	4c	5c	6c	7c	8c	9c
D (iv)	— macro social structure	1d	2d	3d	4c	5c	6c	7c	8c	9c

want of available evidence.

The third reason for dissatisfaction with the table is uncertainty about the orders and types of change, as to their relative pace, their distinctiveness, their completeness and their logic. It is not evident, for example, that the logic of political change is the same as that of economic, constitutional or normative change. Nor is it necessary that there should be a single hierarchy of orders and types of paces of social change. More than one hierarchy may be possible, as has been shown to be the case with social status. Moreover, arranging the categories into descending orders of pace, as we have done, and allotting them Roman numerals within each order, runs the risk of asserting that a series of nominal distinctions constitute steps on a continuous ordinal or even an interval scale.[1] Let it be insisted that the orders, alphabetically designated, are indeed assumed to have the character of a quasi-numerical hierarchy, though the question of the intervals between them has to be left open. The types or categories within the orders are also assumed to have something of an ordinal relationship with each other, at least in the case of the first and the last category with each order. But the relationship between neighbouring types of categories is assumed to be for the most part nominal.

The creation of nine modes of social change as columns in the matrix was unwillingly undertaken and was perhaps unnecessary at such length. But the detailed discussion of the first, the political row or category, in comparative terms under each of the nine column headings will indicate, I hope, that these modes can help us with the problems of logic to some extent. They are intended to illuminate the issue of paces of change by drawing attention to the manner in which change goes forward. Nevertheless, as has already been said, a great deal of guess-work has been necessary as well as much arbitrary and uncomfortable decision making. This is especially so in indicating for each cell, with only five left undecided, distinctions between change which are always, occasionally (or potentially), or never of the character given.

These rather burdensome complications must not prevent us from appreciating the force of the thesis being advanced, which is contained in the unitary hierarchy of paces at the left of Table 2.2. The claim is being made that time itself differs from social activity to social activity. It is time itself which is held to be variable because, as so much of the discussion of the phenomenon goes to show, time is not an absolute but a relative thing, relative to change and therefore different when the pace of change differs. It seems to me to be quite justifiable to conceive of political time, that is to say the order of experienced duration in which political events succeed each other and political developments proceed, as being different from demographic time, for example, religious time, normative time or social structural time.

I must stress the social, collective character of the times and paces of change which are being discussed. Personal time or times are not included in the table, although the pace of change in personal activities will enter into the discussion. Moreover, the definitions of the social activities regarded as distinct for analytic purposes cannot be entered into.[2] Politics, for example, is assumed to be recognisable as such by every reader. It is presumed that it is legitimate to compare pace of change within the political domain with pace of change in the other domains listed, without entering into the question of how change or paces of change can be measured. These may be simplistic assumptions, and provoke disagreement, but the object at this earliest stage is to invite discussion.

Since political events and developments are supposed in this way to happen faster than demographic, religious, normative or social structural developments, the existence of a hierarchy or even hierarchies of social times entailed by differing orders of pace of change seem inevitable. Hence the title of this essay, *Social structural time*, social structural time being claimed to proceed at the slowest pace in the hierarchy, as opposed to political time which is claimed to proceed at the fastest. The obscurity, even the confusion, of language here is due to the universal disposition to refer to time and change, or pace of change, in interchangeable ways. People may say that events were fast, or they may say that the time went by quickly, but they seem to mean the same thing by the two statements. Because the social structure and its changes are and have to be the major objects of scientific study, social structural time is correctly placed in the titular position at the beginning of this essay.

Given that time differs from category to category of social activity, mistakes in attributing to particular categories time changes which belong with other categories are not unlikely. The traditions of historical inquiry being so closely bound up with political events and change, and political time being, as will be suggested, identical in pace to personal time, it is understandable that historians should often tend to make the error of order which attributes the political time scale to other categories and to other activities. The most serious misapprehension is that of attributing the time-scale allotted here to political change, to social structural change. This is to assign the fastest pace of social change to a category in which pace of change is claimed to be slowest. It is to assume in fact that the social structure itself can shift as quickly as opinions or as fashions in clothes.

The occasion of the present exercise was the attribution of the time pace of political change (type A(i)) to micro social structural change, with particular reference to the family (type D(iii)).[3] Such ill-advised attributions are the easiest to pick out and to correct, and when the preoccupation with politics above all things gives way to the attitude of the historical sociologist who expects time and pace to differ from activity to activity, they will no doubt cease to be made. It is not these gross confusions, but the differing times and pace-scales between categories rather closer together in the suggested hierarchy which pose the serious difficulties.

What have we to go on, for example, other than intuition or common sense in determining whether changes in opinion (type A(iii)) in fact go on more rapidly than intellectual changes (type B(v)) as is assumed in the table? How could we decide on the more complex question of whether any one, or two, or even all of the processes of change labelled economic (B(i)), communications (B(ii)), or technical (B(iii)) could swap their orders of pace of change with that labelled fashion (A(ii)) for example? Both questions of course again raise difficulties about the discreteness of the orders and categories of change we are trying to anatomise.

When political change and social structural change were being contrasted, the pace of events was cited in the matter of politics, but not in that of social structure. This is because there is no such thing as a social structural event. In our table we class all the types belonging to our first order of pace as event-like, and the possibility of some types having an event-like character is allowed for as far down in the pace hierarchy as type C(iii), changes in religion. But only the five categories placed in our first order, that of the fastest-paced social times and social changes are claimed to be invariably event-dominated and also event-expressive. The categories classed as very slow, on the other hand, are regarded as event-irrelevant. Events cannot last for more than a brief period or interlude, whatever the time-scale, so it is quite logical that affirmative signs in column one get fewer and finally disappear as we descend the hierarchy. If there can be no event in any of the social structural categories, there can be no narrative history for these categories, and 'telling the story' is a dubious undertaking for any social time slower than the fast.

The fast pace and event-dominated character of politics gave rise some three centuries ago to the metaphorical expression revolution, a sudden reversal of direction, policy or regime quite to be expected in political life. Nothing indicates more clearly the now universal tendency to attribute political time, the political pace of change, to other times and paces than the frequent use of the expression revolution for change in every one of the 19 categories in our list. Thus we are assured of the frequent occurrence of revolutions in economics, in aesthetics, in techniques, in science, in domestic relations and so on. It sometimes seems that no change is worth discussing unless it can be called revolutionary.

From the sudden and violent political events of the recent and fairly recent past, events wholly or partially imbued with the revolutionist ideology, revolutionists have created a paradigm of the most significant type of change in the social fabric. The final reference of all other changes which are now so freely called revolutionary seems to be to this model. In terms of the categories and modes of our matrix, the revolution paradigm denotes a change usually brought about by insurrection (type A(v), mode 1a, 1b) but occurring in category D(iv), macrostructural change, at catastrophic speed. Macrostructural change always induces change in other categories because it has the character of mode 9c. Its other modes in category D(iv) are: 1a, event dominated; 2, society wide; 3a, necessarily violent; 4c, unique; 5c, linear; 6c, irreversible; 7a, capable of acceleration; 8c, transformative.

Such a phenomenon would evidently violate the provisions of our matrix in a number of directions, ordinal as well as modal, and must be classed as irreconcilable with it. Revolution of this description is inconsistent with the hierarchy because it fuses together two types of pace of change belonging to different orders, the fastest and the slowest. It attributes an event-like character to macrostructural change and contradicts the array of descriptive modes in other ways, confused as this array becomes when two categories are mingled in this way. Those who have come to use the noun 'revolution' or the adjective 'revolutionary' so freely cannot all be supposed to contemplate such implications. Their usual object is to exploit the powerful resonances of these words, to employ them as intensifiers. Nevertheless if the general case we are arguing is at all convincing, the revolutionist habit of mind must be a serious obstacle in the way of apprehending the nature of social change and changes, and even of the nature of society itself.[4]

Let us proceed to the further modes of political change as set forth in the matrix. Political change is judged to be inevitably elite-dominated (2b), occasionally violent (3c), always repetitive in character (4a), to have cyclical tendencies (5b), to be usually reversible (6b), never to be variable in its pace, its very fast pace (7c), never to be transformative (8a) and only occasionally likely to lead to change across categories (9b).

This last claim seems acceptable enough, though the politicians themselves and those with strongly held political convictions might demur, since both may wish to suppose that politics is sovereign over society in every respect. An important reason why the political system and political action cannot be of this character is an implication of the indication in the table that political change can be event-expressive as well as event-like.

Event-expressive is a phrase chosen for the want of a better, and is intended to capture the capacity of the political system to convert occurrences, tendencies, stresses, developments throughout the body politic into political events. Thus the conflict on the British coalfields in 1984 and 1985 became a series of political events, taking place at the political pace. These political events occurred both at Westminster and in the mining areas and occasionally verged on attempted insurrectionary change (A(v)), violent by definition, and being violent, necessarily fast. But the economic and technical changes expressed in these insurrectionary interludes, like those expressed in the exercise of military force (A(iv)), went on at their own slower speed. They were undoubtedly affected by political, military and insurrectionary events, actions and changes, but not assimilated to their times, their paces of change. If, as the striking miners may have believed, an alteration in relations of production was occurring, it proceeded at almost the slowest pace of all. In this way our matrix enables us to analyse what may happen in a manner alternative to that proposed in the revolutionist tradition.

Every one of the categories classed as fast paced is marked in the table as event-expressive as well as event-dominated. But the manner in which change in fashion expresses events not themselves to do with apparel, or furniture, or the design of cars and so on, is clearly very different from the way in which political change expresses such things, and there is an even greater difference when it comes to change in opinion. We may remark that the character of being event-expressive demonstrates that our categories must not be taken as being more than partially auto-generative, that is bringing about change from within themselves at their characteristic paces. Each category is clearly liable to be affected by change in other categories, a fact which bears on their ontological status.[5] But we can go no further in the directions here, and to try to tackle every issue of this kind raised by the 171 cells in the matrix would be a tedious matter, and might soon become unrewarding. There are still a number of remarks to be made about the political category and its allotted modes if the exposition is to be reasonably adequate.

The second mode, elite-associated, is a comparatively simple one from the point of view of time. The fact that all except the fifth of the categories in the fast-paced order are marked elite-dominated, and two of those in the medium-paced order, the intellectual (category B(v)) and the aesthetic (category B(vi)) is of some significance. Pace of change amongst the elite, the best informed, the most mobile, the most reactive segment of the community, is certainly likely to be fast. It is a commonsense observation once again that members of the elite do find the reactions of those lower in the social scale to be slower than theirs, and a case might be made out for the time characteristics of ruling classes, nobilities, bourgeoisies, highly educated minorities, being different from the time characteristics of peasantries, blue-collar communities and proletariats. Such a suggestion might in its turn be criticised as a matter of opinion only, elite opinion of course, and the implication of the column entry (2b) under the elite heading for category A(i), that politics is always elite-dominated, might be resisted as well, along with the same implication for categories A(iii), B(v) and B(vi).

The repetitive nature of political change (A(i) mode 4a) and its reversibility (6b), together with its occasional tendency to be cyclical (5b),[6] require a little comment. Repetitiveness and reversibility are likely to be associated with fast- rather than slow-paced change, and it will be seen that every one of the categories in our fourth order, the very slow-paced, is marked as unrepetitive and irreversible. It does not follow from this that no irreversible social change can happen at any pace except the slowest. Economic, demographic and technical, intellectual and aesthetic change may not go on in political time, and it is a frequent error to suppose that they do. But they are often quite rapid and can be irreversible.[7]

Political time is repetitive because of the perpetual recurrence of near identical events, occasions, issues. Social events can scarcely repeat themselves in the same way as physical or even personal events repeat themselves, and it is not surprising to find the judgment 'necessarily repetitive' confined to the first two categories of the table. For this reason, that social times are not interchangeable with personal time, in spite of the fact that personal time is so deeply influenced by social times and proceeds, as I believe, at the same pace, the heading cyclical is less salient in the matrix than might be expected. The circadian and annual rhythms could be claimed to be social as well as individual. We all go to work and come back home together, for example; take our holidays at the same time of the year, send our children to school and celebrate festivals on the same days. These activities might perhaps be described as coincidentally rather than as necessarily social. But the presence of such tendencies in economic change, demographic change, and their possible presence in change of opinion or aesthetics, could be said to be a factor in ensuring that change in these categories is faster than in categories D(i) to D(iv), the most gradual of all.

These circumstances bring to notice an architectural feature of the matrix which is of importance to the pace hierarchy. One of the functions of such institutions as the family, or in certain societies, especially at times earlier than our own, of the church, is the containment of repetitive and cyclical events and change, or providing the stable framework in which such things can go forward. This was remarked upon when the quite unexpected rate of turnover of population in pre-industrial English communities was discovered, nearly two-thirds of the whole population removed by emigration and death being replaced by immigrants and births within a span of twelve years. 'An unchanging, unchangeable social structure may well be essential to a swiftly changing population',[8] which underlines the absurdity of attributing the pace and character of political change to the social structure in which political events and changes go forward, as well as changes in fashion, in opinion, in demography, even in religion. The change-containing character of our structural categories, change of an individual as well as of a social character, is of great importance to the subject.

If there are repetitions or cycles in the very slow-paced social times in which changes in norms or in the family take place, then they must, it seems, have a very long wave-length, like that of the aeons which mark the one great temporal cycle of St Augustine's *City of God*. We may, however, have to revise this judgment somewhat before we reach the end of our discussion of social structural time. Markings for political change in the three final columns are still left for discussion, illustratively and comparatively. These modes are pace variable (7); transformative (8); and causative across categories (9).

Only the last of these columns is marked with as much as a potential affirmative for politics, and the uncertain capacity of political change to entrain other forms of social change has already been remarked upon. It does not seem possible that the pace of politics, political time, can get faster or slower, or that entirely political change can be stadially exclusive, that is replacing formerly existing states so completely that the change is transformatory. For examples of these accelerative, transformative, and widely causative modes, we shall have to turn to other time categories.

But there is a general characteristic of political time which first has to be considered. This is its function, which has already been hinted at, its function as a link between the individual and the collective, the personal time dimension and social time dimension. It could be that we expect all social time or times to be the same as political time because political time elapses at the same pace as personal time, or is felt to do so. Other contributors to this volume, and Michael Young's *The Metronomic Society* (1988), indicate that the unitary character of personal time is itself illusory, that time is different for a person at different stages of the life course, at different times of the year, even at different hours of the day. Nevertheless it might still be true that social events and change are readily appreciated only if they are thought to go on at the political pace, which is also the personal pace, uncritically assumed to be unitary. Indeed, though this takes us beyond the bounded scope of our discussion, the role of the political could be said to be that of making collective and social into personal in every direction, converting collective developments, frictions, tendencies into conflict, event-dominated conflict, between living persons.

If these are justifiable suggestions, then the historian who assimilates time paces of a different order to the time pace of politics is indeed substituting the dogmatic for the critical in analysing change or in relating events. Some support for the view that the only readily acceptable type and pace of change has to go on at the political, the personal pace, could be found in the reactions to change of older people. Slower-paced changes are often not perceived by them, or if perceived, not allowed for in conduct, and always resented as confusing, especially changes in the very slowest categories, religious, normative, familial. Faster-paced changes, proceeding on the political time-scale, do not seem to have such an effect.

Older people, moreover, are those most likely to assert that the pace of change, all social or collective change with the exception of the political, has now speeded up to such an extent as to leave them behind. Since older people are in an excellent position to judge from their own experience whether social changes have been accelerating, and are accelerating still further, their evidence has to be given due weight, although it is possible that one of the necessary features of becoming old is to feel that everything is now changing too fast to keep up with.[9] But the elderly are not the only contemporaries who observe a speeding up of social change, in many or in all directions. If it is overall change of pace which is at issue, uniform throughout our categories, then it would be occurring in defiance of the provisions of our matrix, but might lead to bewilderment all the same. The vague and indiscriminate character of claims like this, and their sensationalism, makes against their being taken seriously.[10] But though we have not allowed political time to be capable of acceleration, we have supposed that acceleration is always or sometimes possible in a clear majority (12) of our 19 categories. These possibilities are of evident interest for the study of social structural time.

They affect our hierarchy generally because a category capable of being speeded up could clearly get to be faster in pace than the other with which it is grouped, and so become a candidate for joining an order above its own. Such eventualities have already been discussed in relation to some of the categories in the medium-paced class in the table, and we turn to those numbered B(ii), B(iii) and B(iv), to do with communications, technics and demography, to proceed with the discussion.

As the entries in the matrix now stand each is marked as

occasionally or potentially capable of acceleration. It might be argued, however, that these judgments should be modified, at least for highly industrialised developed countries like our own. There the speed of change in communication of all kinds, the transportation of persons as well as the transmission of messages, and the speed of change in technical and in demographic directions, all appear to be continuously getting faster, though they only occasionally did so and do so in societies of other kinds. If this were accepted, it might be held to imply that for us these categories of pace and change are now accelerating, and that the phenomena to which they refer are becoming transformative in their operations, causative across categories. Lines B(ii), B(iii) and B(iv) would, therefore, have to come to be marked 7a, 8a, 9b in the modal columns.

The effect of modifications of this kind on our table, particularly those specified, turns out to be quite radical. Indeed, it could be said to place the whole array in danger of flux. The project of working out a hierarchy of social times in relation to paces of change becomes one more dimension more complex as a whole. Different hierarchies would have to be worked out for different social structures under these circumstances, primarily industrial and non-industrial social structure, and a yet further matrix worked out for making comparison between these differing tables. However determined we are to get at the truth about social structural time, we might judge that such would be an impossible project. Its daunting complexity and intricacy, however, underlines one of the highly significant properties of the sociology of time as a study. For we are relentlessly required to recognise the apparently unending complication of social phenomena, especially when they are, as they always have to be, considered as dynamic. Our human scale and human pace, our foreshortenings and approximations, the commonsense manner in which we find our way around the social universe, fall short of a conceptual understanding by an enormous margin.[11]

We should not, therefore, shrink from accepting these possible alterations because of the effects on our painstakingly-worked-out matrix, or because of the complications which would ensue. The amendments to our table would imply that these three categories, quite apart from being likely to mount in the hierarchy of time speed, would have the characteristics at present reserved to very slow-paced changes in the table. This is so because each would become transformative and each would be affecting all other categories in a causative way, and this in perpetuity. We have tried to avoid complications arising from the interaction between our categories, since our object is decidedly not an attempt at total social description. But this ceases to be possible when we consider accelerative potentialities.

The reader is left to decide how persuasive such a revised picture, or any version of it, really is, and how justifiable the impression so many people have – for this once more is all we have to go on, personal impression – that social time is now speedier than ever before, and change more definitive. It will be noted that in the altered hierarchy it would still be the case that social structural change would be going on at a slower pace than the others in our series of classes, and that the attribution to it of anything like the political speed would remain an impossibility.

Nevertheless, there is a particular effect of interaction between categories, that between demographic and familial change, which might convince us that the social structure itself could be altering in our own day at a rate which is on a level with several of those previously classed as considerably faster. The shrinkage of the numbers of children and numbers of kinsfolk due to the drastic fall in births; the sudden ageing of the population; the transformation of procreative life due among other things to almost entirely efficient contraception; the radical modification of the life-course consequent upon these and other developments, including normative developments; all these things have brought about a familial, that is microstructural, change at a pace which might now be thought no longer realistic to regard as very slow. There is plenty of evidence to suggest that the family is changing in a country like Great Britain at an entirely unprecedented speed with extensive effects elsewhere in society, and there is no lack of voices proclaiming that unless this process slows down something like general social confusion is on the way.

Pressing the argument to this point, however, creates an air of unrealism, at least in the mind of the writer of this earliest attempt at an historical contribution to *The Rhythms of Society*. The complexity of the subject, the difficulty of conceptualisation, the uncertainty of the categories, the want of information about pace and relative pace render clear and convincing decisions almost impossible. There is, moreover, a phenomenon well known to theorists of history which might dissuade us from accepting any argument which puts our own present into a peculiarly difficult and disadvantaged position in comparison with all past presents.

There has always existed a disposition for contemporaries to believe that the world as it is at the period when they happen to inhabit it is a victim of the whole historical process. Though they have occasionally congratulated themselves on their greater wealth and their enlightenment, their superior capacity for survival, even their moral superiority to the men and women of the past, all earlier commentators on what was 'now' for them have tended to believe that that 'now' made them uniquely badly off. It is correct, I think, to resist this tendency to be sorry for ourselves about our position in comparison with earlier generation, and to reject any attempt to assess our relationship to generations yet to come. Although it is difficult to deny that the acceleration of changes and of times which we have been discussing may be proceeding in some form or other, it is my own view that we cannot yet go very much further than the suggestions made here in analysing what is going forward. Nevertheless our ignorance and incapacity should not be taken as grounds for believing the worst.

The fact remains that social time for the historical sociologist is plural not singular; that the paces of change which determine that there shall be a plurality of social times differ from category to category of social activity; that it is our duty to get the order of those paces and social times as correct as we possibly can, and always bear in mind that acceleration may have to be allowed for. This duty may baffle us by its complexity and its difficulty, but it is not one which we should try to escape.

NOTES AND REFERENCES

1 For the nominal, ordinal and interval scales, see any textbook of social statistics, e.g. H.M. Blalock *Social Statistics*, 1979, (revised second edition), New York, London, McGraw Hill, pp. 15–19.

2 Their ontological status is thus left undefined, but it must be obvious that they are to be regarded as nominal entities rather than as realist ones.

3 See P. Laslett, 'The duties of the family historian and the character of the history of the family', *Journal of Family History*, vol. 12 (1–2) (1987), pp. 263–84. The reference there is to the work on familial subjects of historians whose earlier writings have been on political history. Such writers tend to take to their subject matter the expectation that there will be familial changes between periods beginning and ending at politically defined marking points (centuries, reigns, intervals between wars or revolutions, and so on) all occurring at the political pace. Alternatively it might be assumed by historians of religion or intellectual life that there must have existed for example a 'Reformation family' which superseded the 'pre-Reformation family' at the religious pace: or an 'Enlightenment family' which came into being at the intellectual pace, and so on.

4 See Chapter 6, 'Social change and revolution in the traditional world' of P. Laslett, *The World We Have Lost Further Explored*, London, Methuen, 1983. Even more widespread is the use of the word 'crisis', to emphasise or to exaggerate the significance of particular interludes in changes of every description. Crisis is also a political concept and belongs, like revolution, to the fastest order of social time, though not exclusively. There can be occasional crises in economic or in demographic change, for example.

5 See above, note 2, and the final paragraph of the text.

6 'Repetitive' and 'cyclical' are perhaps the least discrete of the modes in the matrix. Their distinctness however, can be illustrated from the services performed daily by monks in a cloister. These services make up a diurnal routine which is cyclical, but the routine itself is repetitive in respect of days.

7 It might be supposed that B(vii), institutional change, also classified as medium-paced rather than slow or very slow, would have to be marked as necessarily irreversible. Everyone who has served in a bureaucracy, however, finds that its arrangements are much less permanent than might be

expected. Each new occupant of a position of power tends to rearrange the chain of command and the distribution of functions as much as possible. Institutional time is certainly relatively fast. Totally new institutions require social structural change, and such change accordingly goes on in social structural time.

8 P. Laslett, *Family Life and Illicit Love in Earlier Generations*, Cambridge, Cambridge University Press, 1977, p. 79. [reprinting of a study first published in 1964.] Such rapid turnover, involving repetitive, sometimes cyclical change of residence especially on the part of the substantial proportion of young people who were in service (also coincidentally rather than necessarily social), is now recognised as a commonplace in Western social history. It is conjectured that the family life of our ancestors was far less stable, less variable, less tedious than family life is for us. But the family itself was necessarily rigid in its structure nevertheless, just as our own schools and universities have to be rigid because of the substantial annual turnover in their populations.

9 For some consideration of this and related issues (for example, that the population has changed its age composition so rapidly that recognition of the social consequence is still to come) see P. Laslett, *Britain! Be Your Age!* in course of composition. Evidence from a remoter past might resolve this issue, but no systematic search has been done to my knowledge to discover whether our ancestors ever complained of change going on during their lifetimes so quickly as to be unacceptable.

10 The outstanding example here is the book by Alvin Toffler, with the title *Future Shock*, London, Bodley head, 1970, the superficiality of which has apparently only recently come to be recognised by social scientists.

11 Perhaps social scientists stand in no need of being reminded of these facts, but another way of getting into the position described above is to attempt to imitate social processes by computer, microsimulation as it is called. See K.W. Wachter and E.A. Hammel, 'The genesis of experimental history', in L. Bonfield, R. Smith and K. Wrightson (eds) *The World We Have Gained*, Oxford, Blackwell, 1986, pp. 389–400.

CHAPTER 3

THE INTELLECTUAL HISTORY OF LONG WAVES

Peter Hall

If the social sciences were a nineteenth-century creation, then social scientists have been studying cycles almost as long as they have studied anything. Particularly this was true of the capitalist economy. Say, Ricardo and Malthus were arguing about the phenomenon of 'gluts', or depressions, in the capitalist economy before 1820 (Schumpeter, 1954, p. 621); the French economist Clément Juglar produced a general theory of business cycles, backed by empirical time-series analysis, in 1860 (Schumpeter, 1954, pp. 1123–4). But they all had in mind what has come to be known as the 'classic' business cycle, of some eight to ten years' duration: the one that has continued to obsess economists ever since.

Yet in this, there is an oddity. When ordinary people (and even economists, in everyday talk) think about economic depressions, they think about the big ones: the slump of the early 1930s, or – for those that remember it from the history books – the great depression of the 1880s. These, the real crises of the capitalist system, at first had less attention. Perhaps that was because at first there was only one evident example, the great slump at the end of the Napoleonic Wars. So it was not until the early twentieth century that economists began to try to fit these mega-phenomena into their systems. If the shorter business cycle recurred at regular intervals, was there also some regularity in the longer one? Did the shorter cycles systematically nest inside the longer ones? Thus speculation on the long wave began.

Since then, whether or not the long waves actually exist, there have been indisputable long waves in their academic fashionability. But, unlike women's hemlines, these long waves of fashion run counter to the economic trend: upward in the 1930s, down almost to the ground in the 1960s, up again in the 1980s (Mandel, 1980; van Duijn, 1983). It is not difficult to see why. During the booming 1950s and the swinging 1960s, Keynesian macro-economic management seemed to have put the business cycle, in all its manifestations, almost to rest. Only with the recession of the late 1970s did economists begin to take real notice of it again. And then, it was almost inevitable that someone should dust off the works of Nikolai Kondratieff and Joseph Schumpeter.

KONDRATIEFF AND HIS PREDECESSORS

Ever since Schumpeter's massive two-volume *Business Cycles* (Schumpeter, 1939), the long waves have been named after Kondratieff. Strictly, this is unfair: Schumpeter, the great historian of economic theory, seems totally to have overlooked the earlier contributions of the Dutch economists van Gelderen and De Wolff (van Duijn, 1983, pp. 61–3). Kondratieff had made the same mistake: in his classic paper of 1925, which is the one known to the English world through translation ten years later (Kondratieff, 1935), he had acknowledged that he first learnt of their work after writing his own paper (Kondratieff, 1935, p. 115). But Schumpeter immortalised Kondratieff, which had at least the justice of recognising his work of the 1920s – nearly all still in the original Russian – long after his memory had perished in his homeland.

And, with all recognition to his predecessors, it is the definitive beginning. In his Institute for the Study of Conjuncture (rendered best in English, perhaps, as the Institute for the Study of the Capitalist Economy), Kondratieff was the first to attempt systematic quantitative analysis of the course of economic growth in the major capitalist nations since the first industrial revolution of the eighteenth century. Considering a number of indicators – prices, wages, interest rates, production of basic industrial goods – he concluded that the history was marked by a regular, systematic progression from boom to depression and back to boom again. These swings occurred at 50- to 60-year intervals. Kondratieff dated the upswing of his first wave 1789–90, its downswing 1810–17; his second upswing 1844–51, the downswing 1914–20 (Kondratieff, 1935, pp. 110–11). The cycles were therefore irregular: the first lasted some 60 years, the second only about 47.

Kondratieff was stronger on measurement than on theoretical explanation, a fact for which he was criticised by fellow Soviet economists during the 1920s (van Duijn, 1983, p. 64). And the explanation he did give is interestingly at odds with Schumpeter's later interpretation: it was variation in the rate of renewal of capital goods, which he related to the price of loan capital. True, he identifies a swarm of new innovation as one feature of each upswing (Kondratieff, 1935, p. 112); but he refuses to accept this as the basic explanation:

> Changes in the technique of production presume (1) that the relevant scientific-technical discoveries and inventions have been made, and (2) that it is *economically* possible to use them . . . Scientific-technical inventions in themselves . . . are insufficient to bring about a real change in the technique of production. They can remain ineffective so long as economic conditions favorable to their application are absent . . . the development of technique itself is part of the rhythm of the long waves (Kondratieff, 1935, p. 112)

This in fact provides the other half of the mechanism he needs in order to demonstrate the automatic, reciprocating character of the process. At a certain point in the capital-replacement cycle, a bunch of available inventions is ready to be turned into commercially useful innovations, awaiting only the injection of capital occasioned by lowered profit margins on existing investment opportunities.

And this process was endogenous. If he had fully spelt it out, it might have made his theories even more heretical. As it happened, Stalin arrested him on the ground of organising an opposition party and sent him to Siberia, where he disappeared. But certainly, his theory was very dangerous to

Soviet orthodoxy: it both demonstrated the truth of the Marxist thesis that the capitalist world should be racked by periodic crisis, and simultaneously showed that it would always recover from them.

SCHUMPETER STANDS KONDRATIEFF ON HIS HEAD

In doing so, Kondratieff made a huge theoretical jump. And Schumpeter, who as editorial board member had him translated first into German and then into English, well appreciated its momentous importance. For already in 1911, in his *Theory of Economic Development*, Schumpeter had set forth his revolutionary conclusion that the whole edifice of neoclassical economics was deficient: obsessed with the tendency toward equilibrium, it could explain only the relatively small, self-equilibriating adjustments that occurred in this 'circular flow' which went on 'running in channels the same year after year' (Schumpeter, 1961, p. 61); it could do nothing to explain the major shocks to this system, which constituted economic development. These, Schumpeter had then argued, came endogenously, through what he termed – in a celebrated phrase – 'the carrying out of new combinations': producing a new good, devising a new method of production or mode of organisation, finding a new market or source of supply (Schumpeter, 1961, p. 66). The 'carrier-out of new operations' was not the capitalist, but the entrepreneur (Schumpeter, 1961, p. 75, p. 137).

It was this, Schumpeter had already determined in his twenties, that was the cause of booms and depressions:

> the boom ends and the depression begins after the passage of the time which must elapse before the products of the new enterprises can appear on the market. And a new boom succeeds the depression when the process of resorption of the innovations is ended (Schumpeter, 1961, p. 213).

Further, even then he had concluded, such new combinations appeared 'discontinuously in groups or swarms' (Schumpeter, 1961, p. 223); entrepreneurial activity was contagious, and thus investment would increase and capital-goods industries expand.

At this point, though – he was writing about 1907 – Schumpeter was thinking only of relatively short (or 'classical') business cycles, of about ninc ycars' duration, which he later christened for their discoverer, the French economist Clément Juglar (Schumpeter 1954, pp. 1123–4). Juglar's famous aphorism was that 'the only cause of depression is prosperity'; the only problem was that he could not explain what caused the prosperity. Schumpeter, in 1907, thought he had identified it. But it took him nearly another twenty years to hear about Kondratieff and to incorporate him into the definitive restatement of the theory, published in his 1939 *magnum opus*.

Business Cycles argued that there were at least three business cycles: the Kitchen, named for an English economist, some 38 months long; the Juglar of $9^2/_3$ years; and the Kondratieff, which he confidently pronounced to be 57 years long (Schumpeter, 1939, p. 1051). These nested, one inside the other, 'six Juglars to a Kondratieff and three Kitchens to a Juglar – not as an average but in every individual case' (Schumpeter, 1939, pp. 173–4). His starting-dates for the Kondratieffs, which are subtly different from those of Kondratieff himself – 1787, 1842/3 and 1898 – actually represent lengths of 56, 54 and 56 years respectively; though he does say (oddly, in the light of his confidence elsewhere) that these dates are 'tentative' and 'approximate' (Schumpeter, 1939, pp. 170, 252, 303, 397).

Each of these Kondratieffs represented, in effect, a new industrial revolution resulting from the birth of a new group of technologies:

> Historically, the first Kondratieff covered by our material means the industrial revolution, including the protracted process of its absorption. We date it from the eighties of the eighteenth century to 1842. The second stretches over what has been called the age of steam and steel. It runs its course between 1842 and 1897. And the third, the Kondratieff of electricity, chemistry and motors, we date from 1898 on (Schumpeter, 1939, p. 170).

Schumpeter here is specific: the Kondratieff triggers, the

'innovations of relatively long span' (Schumpeter, 1939, p. 167) were new technologies that created new industries. It is important to recognise, though, that – as Schumpeter was at pains to stress – innovations were not inventions: they were commercial applications, and they did require the capitalist to work hand-in-hand with the entrepreneur. He did, Schumpeter argued, because innovation was the sole force that kept the rate of interest above zero (Schumpeter, 1939, p. 124). And the effects on investors were then multiplied by the generation of an atmosphere of optimism about the prospects of the new enterprises, which he termed the Secondary Wave (Schumpeter, 1939, p. 145).

The critical question – as Simon Kuznets asked in his celebrated 1940 review of *Business Cycles*, and as many have asked since – is why exactly the innovations should bunch at the bottom of the cycle. Schumpeter did not provide the answer. Kuznets tried to do so: perhaps, it was because the possibilities arising from (and, therefore, profits arising from) previous innovations were exhausted. This brings us close to the idea of what Raymond Vernon was much later to christen the profit cycle, though he clearly had in mind a shorter cycle approximating to the Juglar (Vernon, 1966): product innovations would bring forth new industries, which would then proceed down a learning curve, with process innovations producing lower unit costs; the result would be massive economies of scale and increasing output, until market saturation was reached. Empirical work on the growth of capitalist economies during the Fourth Kondratieff suggests that just such a process has occurred: in the first upswing, both production and employment increase rapidly and in step; then, as process innovations take hold, there is a period of jobless growth of output; finally, with market saturation the rate of output sharply falls while employment actually contracts (Rothwell and Zegveld, 1981, Rothwell, 1982).

However, this does not quite tease out the precise mechanism. Schumpeter does suggest it: a new bunch of entrepreneurs seizes its opportunity, just as a previous generation begin to suffer from hardening of the innovative arteries: 'we will assume that innovations are always associated with the rise to leadership of New Men' (Schumpeter, 1939, p. 96).

There is a hint here of a generational cycle, but if so one entire generation misses the boat; perhaps this intermediate cohort is responsible for the process innovations. Somewhere in the Schumpeterian account, there is a critical circuit either missing, or at least hidden from view.

That was not the only stricture Kuznets made. In the lifelong empirical work which eventually won him the Nobel Prize, he argued that the Kondratieff cycle was a fiction: the real long wave was a construction cycle of some 20 years' duration, which has come to be known as the Kuznets cycle (van Duijn, 1983, pp. 15–18). Subsequent work has suggested that this cycle was only really evident in the United States in the period 1840–1914, and that in any event, as a capital replacement cycle it could well nest into the Schumpeterian family: two Juglars to a Kuznets, three Kuznetses to a Kondratieff (Forrester, 1976).

Criticism from authorities like Kuznets was not the only or even the main reason why, during the 1950s and 1960s, interest in the Kondratieff waves should die away and why Schumpeter himself should be reduced to the status of a 'footnote economist' (van Duijn, 1983, p. 97). The real cause was the apparent triumph of the Keynesian analysis in the *General Theory*, which Schumpeter himself felt had detracted from the reception of *Business Cycles*. Yet Schumpeter himself was quite specific that Keynesian analysis, like all neo-classical theory, was totally confined to explaining short-run variations around the equilibrium position, equivalent perhaps to the 38-month Kitchen cycle (Schumpeter, 1952, p. 287); as he put it with studied Viennese courtesy, Keynes's account 'abstracted from the essence of the capitalist process' (Schumpeter, 1954, p. 1175). In blunter language, it did nothing to explain the longer-term business cycles, which everyone believed it was all about.

LONG WAVES RETURN FROM LIMBO

Thus Schumpeter, just before his death, had grasped that the whole of Keynesian demand management policies, on which western economic policies were based for thirty years, were in the long run irrelevant. No wonder Keynes said that in the long run we are all dead. By the time the next Kondratieff downswing came, both he and Schumpeter were – long since. But the next generation of economists were very much alive, and it is no wonder that they began to go back to other models. Many went to Marx for his more complete account of capitalist evolution, which related economic change to social and political forces in a way that neo-classical theory never allowed itself to do; political economy, which had been buried since the days of Jevons and Walras and Marshall, now reappeared.

Logically, some of these economists rediscovered Kondratieff; the Belgian marxist, Ernest Mandel, made the long waves a central feature of his *Late Capitalism*, first published in German in 1974 (Mandel, 1975), and later published an extended analysis of them (Mandel, 1980). Mandel interprets the long waves in orthodox Marxian fashion: they result mainly from fluctuations in the average rate of profit, resulting from the rise in the organic composition of capital (the proportion of capital that, in Marxian terminology, goes on what neo-classicists call capital, as against wages). But, like every previous long-wave theorist, he then faces the big hurdle: how to explain the upswing. Mandel's is that each upswing is special: a war, a gold discovery, rearmament. Certainly – and here too he is impeccably orthodox – there is nothing intrinsic and endogenous to the system, which could generate automatic upswing. But if special factors apply, why the apparent regularity? Once again, there is a mystery.

But non-Marxist economists were turning in the same direction. They shared with the Marxists a notion that technological change could no longer be consigned to the black box where neo-classicists – including the Keynesians – had put it (Rosenberg, 1976, 1982). It was logical therefore that when Gerhard Mensch's *Das technologische Patt* appeared in English translation in 1979, it should prove a major agent – almost without doubt, the major agent – in the remarkable rehabilitation of Kondratieff and Schumpeter in the literature of the 1980s (Mensch, 1979). There is an interesting point of linguistic-cultural tradition here: interest in long-wave theory has always remained strong in the

Netherlands and in Germany, where neo-classical economics never achieved quite the same preeminence as in the Anglo-American culture, and where study of Marx and Schumpeter (who regarded himself as building on Marxian theory) remained a mainline intellectual activity.

Mensch's work, in any event, is avowedly Schumpeterian: he sets out to provide the empirical base that *Business Cycles* lacks, by systematically analysing the pattern of invention and innovation from the eighteenth century to the present day. He concludes that indeed, as the theory requires, innovation has occurred in a sharply bunched pattern. It peaked sharply in four so-called radical years, 1764, 1825, 1886 and 1935: each between eleven and seventeen years before the start of a Kondratieff downswing. He further shows that the basic inventions, from which the innovations stem, do not peak in this way; they show only a weak wave-like motion. So, he concludes, there is a wagon train effect: inventions have to wait for suitable economic conditions before entrepreneurs arise to convert them into marketable products. And, because the process is so regular – its periodicity, he argues, tends to shorten over time – he claims to predict the next radical year, which will usher in the Fifth Kondratieff: it is 1989.

Coming at a time when the capitalist economy appeared to be plunging into the worst depression since the 1930s, it could hardly fail to make an impact. Just as the first great long waves debate represented the economic world's reaction to Schumpeter, so the second represented its reaction to Mensch. And the criticism has taken a very similar form: that the chronology is inaccurate because based on poor data and inadequate analysis; and that the causal explanation is wrong.

The weightiest attack by far has come out of the extended research programme of Christopher Freeman and his colleagues at the Science Policy Research Unit at the University of Sussex (Freeman et al., 1982, 1983, 1985). They argue that his data are incomplete and biased, both as to selection of innovations, and as to their dating; more complete analysis, even using the same sources, shows no tendency to swarming around the radical years. More recently, an exhaustive statistical analysis by Solomou concludes that SPRU are right in concluding that innovation

Table 3.1: Datings for the four Kondratieffs

		Extreme values	*Range*	*'Consensus'*
1st	Upswing	1787/1792	5	1790
1st	Downswing	1808/1825	17	1817
2nd	Upswing	1842/1851	9	1846
2nd	Downswing	1869/1875	6	1873
3rd	Upswing	1890/1900	10	1895
3rd	Downswing	1913/1925	12	1920
4th	Upswing	1935/1948	13	1948
4th	Downswing	1951/1974	23	1974

Source: van Duijn, 1983, p. 163

flows are not subject to long-wave fluctuations (Solomou, 1986, p. 111).

The SPRU group also conclude, with Mandel, that though long waves exist they have no single explanation; each one has a unique origin. And other observers have agreed with them on this. Rosenberg and Frischtak, for instance, suggest that upswings result in part from exogenous forces, in part from the revival of older technologies; thus recovery in the late 1930s came through rearmament plus the revival of the automobile industry (Rosenberg and Frischtak, 1984). But – once again – others disagree. The Massachusetts Institute of Technology group under Jay Forrester conclude that the long wave is generated endogenously, and does not depend on random shocks such as gold discoveries (Forrester, 1976; Sterman, 1985). They conclude that upswings have two causes: first, changes in expectations, whereby firms amplify changes in demand, overproducing in response to them; secondly, further amplification through linkages between firms. So the MIT group follow Kondratieff rather than Schumpeter: instead of an innovation-based theory of long waves, they have a long-wave-based theory of innovation. In the downturn, not only does old capital depreciate, but old companies based on old technologies go bankrupt and traditional methods cease to be sacrosanct (Forrester, 1976;

Table 3.2: Major innovations, by country, by Kondratieff phases

	UK	*Germany*	*France*	*USA*	*Other*	*Total*
1st K Upswing 1787–1812			1			1
1st K Downswing 1813–1842	8	1	3	2		13
2nd K Upswing 1843–1868		4	2	6	3	15
2nd K Downswing 1869–1897	6	7	1	16	2	32
3rd K Upswing 1898–1923	3	5	1	13	1	23
3rd K Downswing 1924–1953	6	13	1	28	3	51
4th K Upswing 1954–1979	3	2		12	2	19

Source: van Duijn, 1983. Datings are Schumpeter's for 1st–3rd Kondratieff; thence the author's, following his chronology. Only innovations ascribed to a single country are listed.

Sterman, 1985). Actually, Schumpeter did say all this; perhaps the debate between Harvard (1939) and MIT (1985) is somewhat metaphysical in character. For the interaction between the long wave and the innovation rate is a reciprocal one: both cause each other.

Meanwhile, those who attack Mensch on statistical grounds are open to the charge that there is no real agreement on the timing of the long waves; indeed, the disagreement widens as we come closer to the present (Table 3.1). Van Duijn, who has done the closest recent work to establish a precise chronology based on recent international data on economic growth rates, concludes that on this basis innovations do vary systematically. And Table 3.2, which is based on his listings but on Schumpeter's chronology projected into the Fourth Kondratieff, strikingly shows this variation. The debate continues.

OPENING UP THE DEBATE

Meanwhile, some of the protagonists have opened up the

debate in an interesting way. Both Christopher Freeman and his colleagues, and the Venezuelan economist Carlota Perez, have suggested that the whole concept of innovation needs to be extended to include not merely technological, but also institutional and social changes. Perez in particular argues that each long wave has a particular technological style, a 'paradigm' for the most efficient organisation of production. In the upswings, there is a fit between this paradigm and the socio-institutional framework; but in the downswing, a new technological factor inevitably brings a disjuncture with the existing socio-institutional system. The resulting adjustment will be long and difficult, and will be marked by a crisis, with considerable experimentation and conflict between existing vested interest and new forces. Thus, oddly, the new paradigm will emerge decades before it actually dominates: Fordism, basis of the mass production paradigm, appeared in the 1910s but became dominant only after World War Two, when major institutional changes – Keynesianism, consumer credit, scientific management, labour relations – provided the right framework (Perez, 1983).

Again, there are echoes of this in Schumpeter: he too was very conscious of the role of social innovation. But the precise triggering mechanism, once again, is elusive. The new paradigm comes in the upswing, and appears to correspond to a process innovation in the product cycle. This is logical enough: it results from the attempt by a leading, highly innovative producer to cut costs and gain market advantage. But what then brings the crisis, which causes the new paradigm eventually to win out? A new technology; but why? Again, its appearance is a mystery. What has put it there? Does it somehow drop into place of its own accord, as some allege Schumpeter said? Or is it triggered by falling profits, as the Marxists should say, but some (Mandel) do not? The answer is still not clear.

If a spiritualist medium could reach Schumpeter he might answer something like this. Generations grow up and react to the environment – technical, social, cultural – of their youth. Some, the brightest, react against it; they have new ideas. (Recall Schumpeter's strange insistence that no one had a new idea after the age of thirty.) Some few of these, in turn, manage to turn these ideas into reality: they are the entrepreneurs. This condition is a permanent human one: Roman poets wrote about it two thousand years ago. But capitalism's unique dynamic harnesses and uses it: Brunel, Edison, Ford, Jobs are its legendary heroes.

Capitalism does this because of its inherent logic, which Marx was the first to identify but which Schumpeter reinterpreted. Because of the accumulation of capital, by an iron arithmetic rule, the profit rate must fall. And further, since this results in substitution of capital for labour (in Marxist jargon, a rise in the organic composition of capital), unemployment must rise. So capitalism must, by its own logic – its own success, in fact – come to crisis.

Schumpeter modified Marx in three crucial ways. First, the driving force of capitalism is the entrepreneur, who is different from the capitalist. Second, therefore, though over time the entrepreneur's profit must fall, so must the capitalist's return, which is the interest on the money he lends the entrepreneur. (This is why Schumpeter was so insistent that in equilibrium the rate of interest would be zero: a proposition that caused his contemporaries to think him mad.) Thirdly, the only way that capitalists can escape from this paradox is to find entrepreneurs who can supply them with innovations, which will give them temporary monopoly rents. Later, as the innovations diffuse, these will again erode and disappear; and the reciprocating mechanism will again come into play.

This is the brilliance of the Schumpeterian structure. It has the inevitable logic of Marx, but it explains why each time, capitalism overcomes its crisis. Further, it is quite capable of bearing the whole load which subsequent interpretation has put upon it. The key innovations are so great that they do require enormous socio-cultural transformations; the substructure of economic forces does drive the superstructure of social relationships, transforming capitalism from one form into its successor: periodically, as the young Marx put it, the integument of capitalism bursts asunder – yet each time, a new nucellus is found within. And, though the reciprocating mechanism itself has a monotonously regular rhythm, the necessary adjustments are so complex that they are far from regular or predictable. Some societies, some nations, may fail entirely to make them: their institutional structures or inherited cultures are simply too ossified. For them, Schumpeter's process of constant creative destruction may not be creative at all: the logic of capitalism has no mercy. The relentless decline in British innovative performance, which emerges from Table 3.2, tells its own story: this is a country that is failing to make the necessary leap from one Kondratieff to the next.

There was, for Schumpeter, one final irony: in this course of adaptation, capitalism would inevitably destroy itself. In his most popular work, *Capitalism, Socialism and Democracy*, he posed and answered his famous rhetorical question: 'Can capitalism survive? No. I do not think it can' (Schumpeter, 1942). This was not because of the immiseration of the proletariat and resultant revolution, as Marx had once supposed. Among his many uncanny predictions, Schumpeter correctly forecast in 1942 that the long-run performance of the capitalist economy would be sufficient to allow it to overcome the next Kondratieff crisis with relative ease, through generous payments to the unemployed. His only error was to underestimate the performance of capitalism in the Fourth Kondratieff, which was almost certainly the best in its history.

No: capitalism would not survive, because the inevitable changes in the system would turn it into something else. Large, bureaucratised corporations would increasingly dominate the economic landscape. In them, the process of innovation would be regularised and systematised. The path to socialism would be eased, because finally it would be irrelevant whether these huge monoliths were public or private corporations. Here, Schumpeter was making exactly the same argument as the early American trotskyist writer James Burnham, whose book *The Managerial Revolution* caused such a sensation on its appearance in 1941 (Burnham, 1941).

We cannot say how a rematerialised Schumpeter would comment on the economic history of the past forty years; the growth of the multinationals, the new international division of labour, the vast expansion of paper capitalism and the hollow corporation. But we can guess that most of these trends would have been pure grist to his mill. For they represent an increasing divergence between the two arms of the reciprocating mechanism: capital is ever more divorced from entrepreneurship as he understood it. The managerialist corporation increasingly drives itself, while capital increasingly flies off into the speculative world of leveraged buyouts, junk bonds and greenmailing.

Or, perhaps, Schumpeter would say that this last is a typical warning sign of the recession phase of the Kondratieff: in what he called the Secondary Wave, capital – desperately seeking profitable outlets – turns ever more to speculation. It has all happened before: not the details, of course, but the essential process. It happened at other times in history, always before the great Kondratieff crashes. And, according to Schumpeter's own chronology, the next of these – the coincidence of the downswing phases of the Kitchen, Juglar and Kondratieff – is timed for July 1987: the nadir of the Fourth Kondratieff. That month, the capitalist world had better go around on tiptoe, hoping that the old master was wrong after all. But, given the uncanny rightness of much else that he wrote half a century ago, it is not a good bet.*

REFERENCES

Burnham, J. (1941), *The Managerial Revolution*, New York, Putnam.

Duijn, J.J. van (1983), *The Long Wave in Economic Life*, London, George Allen & Unwin.

Forrester, J.W. (1976), 'Business structure, economic cycles and national policy', *Futures*, vol. 8, pp. 195–214.

Freeman, C. (ed.) (1983), *Long Waves in the World Economy*, London, Pinter.

Freeman, C. (1985), 'Long waves of economic development', in T. Forester, *The Information Technology Revolution*, Oxford, Blackwell, pp. 602–16.

Freeman, C., Clark, J. and Soete, L. (1982), *Unemployment and Technical Innovation: A Study of Long Waves and Economic Development*, London, Pinter.

* This was written in late 1986, and was based on an earlier draft of 1983. On October 19 1987 the Dow Jones Industrial Index fell 508 points.

Kondratieff, N.D. (1935), 'The long waves in economic life', *The Review of Economic Statistics*, vol. 17, pp. 105–15.
Kondratieff, N.D. (1984), *The Long Wave Cycle*, New York, Richardson & Snyder.
Mandel, E. (1975), *Late Capitalism*, London, New Left Books.
Mandel, E. (1980), *Long Waves of Capitalist Development*, Cambridge, Cambridge University Press.
Mensch, G. (1979), *Stalemate in Technology*, Cambridge, Mass., Ballinger.
Perez, C. (1983), 'Structural change and the assimilation of new technologies in the economic and social systems', *Futures*, vol. 15, pp. 357–75.
Rosenberg, N. (ed.) (1976), *Perspectives on Technology*, Cambridge, Cambridge University Press.
Rosenberg, N. (1982), *Inside the Black Box: Technology and Economics*, Cambridge, Cambridge University Press.
Rosenberg, N. and Frischtak, C. (1984), 'Technological innovation and long waves', *Cambridge Journal of Economics*, vol. 8, pp. 7–24.
Rothwell, R. (1982), 'The role of technology for industrial change: implications for regional policy', *Regional Studies*, vol. 16, pp. 361–9.
Rothwell, R. and Zegveld, W. (1981), *Industrial Innovation and Public Policy: Preparing for the 1980s and 1990s*, London, Frances Pinter.
Schumpeter, J.A. (1939), *Business Cycles*, New York, McGraw Hill [reprinted in 1982, Philadelphia, Porcupine Press].
Schumpeter, J.A. (1942), *Capitalism, Socialism and Democracy*, New York, Harper.
Schumpeter, J.A. (1952), *Ten Great Economists: From Marx to Keynes*, London, Allen & Unwin.
Schumpeter, J.A. (1954), *History of Economic Analysis*, New York, Oxford University Press.
Schumpeter, J.A. (1961 [1911]), *The Theory of Economic Development*, Oxford, Oxford University Press.
Solomou, S. (1986), 'Innovation clusters and Kondratieff long waves in economic growth', *Cambridge Journal of Economics*, vol. 10, pp. 101–12.
Sterman, J.D. (1985), 'An integrated theory of the long wave', *Futures*, vol. 17, pp. 104–31.
Vernon, R. (1966), 'International investment and international trade in the product cycle', *Quarterly Journal of Economics*, vol. 80, pp. 190–207.

CHAPTER 4
VIVOS VOCO:[1] RINGING THE CHANGES IN THE HISTORICAL GEOGRAPHY OF TIME CONSCIOUSNESS

Nigel Thrift

Nearly forty years ago, when teaching in South Wales, I often spent the summer half-holidays between noon and midnight in tracking some small tributary of the Towy to its source in the mountains, and this led me by devious ways through many solitary fields. Over and over again, when the slanting shadows were beginning to show that beautiful countryside in its most beautiful aspect – when those words of Browning's Pompilia came most inevitably home: 'for never, to my mind, was evening yet but was far beautifuller than its day' – over and over again, at these moments, I found myself hailed by some lonely labourer, or by one of some small group, leaning on his hoe and crying to me across the field. It was always the same question: 'What's the time of day?' – the question implicit in that verse of Job: 'As a servant earnestly desireth the shadow, and as an hireling looketh for the reward of his work.' The sunlight was not long enough for me on my half-holiday; it was too long for these labouring men; and the memory of those moments has often given a deeper reality to that other biblical world:

> *Behold, the hire of the labourers who have reaped down your fields, which is of you kept back by fraud, crieth: and the cries of them which have reaped are entered into the ears of the Lord of the Sabaoth . . . Behold, the husbandman waiteth for the precious fruit of the earth, and hath long patience for it, until he receive the early and later rain. Be ye also patient; stablish your*

hearts: for the coming of the Lord draweth nigh. (Coulton, 1926, p.1)

1 INTRODUCTION

This chapter provides an outline of the historical geography of time consciousness in England in the years from 1100 to 1300. The main concern is to represent time consciousness in this part of the medieval period as *an ideology of everyday time practice*, that is as a set of accounts of (or conversations about) temporal practice[2] through which people interpret/construct what 'time' is. Accounts of temporal practice are by and large lost to us for this period, especially for the lower strata of society, so that an indirect approach to their representation is required. Accounts of temporal practice are funnelled through three main channels of discourse upon time which both ensure that the key elements of temporal practice can be the same at many different places and also allow temporal practices to be controlled at a distance. These three channels are, respectively, documents or texts, devices or instruments, and systems of discipline or drill. The historical geography of these three main channels of discourse can be reconstructed for the period from 1100 to 1300, at least in outline. In turn, this reconstruction can be used to speculate on the different accounts of time that were possible in different places in England (Thrift, 1985).

It is important to note three caveats at this point, First, I do not want to claim that the period from 1100 to 1300 is *the* period in which certain aspects of time consciousness in England start to take on a recognisably modern form; even as early as the Anglo-Saxon period the growth of social institutions and a corresponding increase in the sophistication of territorial organisation (typified by hides, renders, etc.) required a practical organisation of time which must have led to a more acute consciousness of time. Rather, I want to suggest that recognisably modern accounts of time can already be found between 1100 and 1300 which, at the very least, give pause to the idea that certain aspects of modern time consciousness are phenomena rooted in the eighteenth and nineteenth centuries. This rather arbitrary form of bracketing will be alien to many historians who tend to start a project with a significant date. Whilst there are clearly many significant dates in the history of time consciousness in and around the period from 1100 to 1300 – for example, the creation of Chancery by the early Plantagenet (Angevin) Kings – I am more concerned here with establishing how in an arbitrarily chosen part of the medieval period, changing customs and practices were already tending to produce modern forms of time consciousness.

Second, in making this examination of changing customs and practices, I will want to move away from those works that posit a single medieval world view of time, often culled from the literary, historical and religious sources of the period. Such studies are quite common, but they are also suspect, not only because of their attempt to create a unified, unitary and non-contextual time consciousness but also because, given the sources they use, they nearly always constitute 'the view from the top'. This distancing means I will have little to say about issues like 'cyclical' versus 'linear' time or medieval theology and eschatology, important though these may be. Instead I will be concerned here with more prosaic changes in the material culture of the period that lead to changes in the ideology of everyday time practice (and so time consciousness) and how these changes were sieved through the grills of social rank (who) and location (where).

Third, I am well aware that the historical account given below may lean towards a technological determinism. This is, partly at least, a function of the evidence that still survives – books and bells leave their trace longer than the day-to-day doings and conversations of the past.

2 AN EXPLORATION OF NEW FORMS OF TIME CONSCIOUSNESS IN MEDIEVAL ENGLAND

It would be possible to draw a picture of England in the period from 1100 to 1300 as inhabited by a people of great temporal innocence and this is a common strategy in the literature. Natural rhythms dictate the pace of life and work and the content of language. Memory of the past is scanty or non-existent – for example, few people have any idea of their calendrical birth date but instead refer to genealogy and important events to locate themselves in time past. Expectation of the future centres on a short life-span and the imminence of the Day of Judgement. Above all, 'in high technology cultures today, everyone lives in a frame of abstract computed time enforced by millions of printed calendars, clocks and watches. In twelfth century England there were no clocks or watches or wall or desk calendars' (Ong, 1982, p. 97). All this is true – to an extent. But that extent is more limited than was once imagined. For the twelfth and thirteenth centuries are being revealed as a time when time consciousness was somewhat more developed than was originally thought. In particular, it is important to stress that just because people have no access to sophisticated time reckoning instruments does not mean that they are unsophisticated about time.

Thus, even in the deepest countryside, the rhythm of the year was never simply the twelve lunar months and the four seasons, for there was always the rich temporal patchwork of the ecclesiastical round (Philip, 1921; Homans, 1941), the key system of time discipline. In the context of the countryside the ecclesiastical calendar (a merging of a calendar of Christ's life with a calendar of saints) had become the centre of a sophisticated ideology of everyday time practice, a storehouse of different but inter-related practices and traditions (Homans, 1941) in which the basic routines of life revolved around the four great pivots of Christmas, Easter, Lammas and Michaelmas. It was a religious calendar, certainly. It was, as well, a secular calendar (insofar as the division had meaning in medieval times) referring to days of work, days on which no work was done, feast days, market days, days for paying rent and so on. It was also a ritual calendar complete with, for example, lucky and unlucky days for holding baptisms and seasons when marriage was forbidden. And, finally, it was an almanac of arable agriculture. The peasant *knew*, with the kind of certainty that we now reserve for principles of natural science, that 'lambs conceived at Michaelmas would be born before Candlemas; that the ploughing should be over by Andrewmas; that ewes should go to tup at St. Luke; that servants were hired at Martinmas and that hay fields should not be grazed for more than a fortnight after Lady Day' (Thomas, 1971, p. 738). Homans (1941, p. 379) summarises the calendar as well as any:

> By the thirteenth century the sequence of the husbandman's feast days had become well adapted to the sequence of the weeks of the year. The greater feasts fell in the interstices between the farming seasons, when one kind of work had been finished and another not yet begun. Thus Christmas came after a Season dominated by the sowing of the winter cornfield had ended and before the sowing of the spring field had begun, Easter at the end of spring sowing, and the wake commonly between harvest and the (autumn) ploughing. This arrangement is the natural one, if only for the reason that men must have leisure for sport and ceremony. As for the lesser feasts, many of them had become linked with a particular kind of farm work. Thus Martinmas was the traditional time for slaughtering, and Candlemas the traditional time for putting meadows in defence. The working of the year and its festivals intermeshed to make a single and stable cycle.

Into this calendar were set the other units of time. There was the month, ruled by the waxing and waning of the moon. There was the week, marked out by the sacred and secular significance of Sunday (Colson, 1926). And, if there was no bell in earshot to mark the passage of the day, there might be the hayward's horn or just a simple stick searched out to stand in the ground and act as a primitive sundial.

As if the sophistication of this basic system of time discipline is not evidence enough, there are other indicators. Certainly, the sense of the past and future imbricated in the accounts of the inhabitants of medieval England would have been much attenuated but it is important to remember that here were people who built cathedrals which required quite exact temporal planning and an idea of a project lasting well into the future.

Thus England in the twelfth and thirteenth centuries was never a bucolic dreamtime. Certainly awareness of the passage of time was still primitive in many places, what Gurevich (1985, p. 143) calls a 'slow-moving, leisurely and protracted affair', but it was not, if this does not seem too great a paradox, unsophisticated. Coulton's (1926) nineteenth-century rural South Wales would not have seemed so different. Further, amongst certain social ranks living in certain places awareness of the passage of time and what this passage signified was being changed by the advent of new texts, devices and systems of discipline that went to make up a new ideology of everyday time practice. The new forms of time consciousness thus engendered had a very definite social geography.

2.1 *Texts*

'When the voice has perished with the man, writing still enlightens posterity.' Thus runs a twelfth-century Winchcombe Abbey cartulary (cited in Clanchy, 1979, p. 202). It points out that texts enable memory to be organised in a way different from that found in oral cultures and, in doing so, they promote a different kind of time consciousness. But, of course, in medieval England texts were only fitfully available, even to those who could read. Ong (1982, pp. 97–8) provides one account of the resultant culture:

> Before writing was deeply interiorised by print, people did not feel themselves situated every moment of their lives in abstract computed time of any sort. It appears unlikely that most persons in medieval . . . western Europe would have been aware of the number of the current calendar year – from the birth of Christ or any other point in the past. Why should they be? Indecision concerning what point to compute from attested the trivialities of the issue. In a culture with no newspapers or other currently dated material to impinge on consciousness, what would be the point for most people in knowing the current calendar year? The abstract calendar number would relate to nothing in real life. Most persons did not know and never even tried to discover in

> what calendar year they had been born . . .
>
> Persons whose world view has been formed by high literacy need to remind themselves that in functionally oral cultures the past is not felt as an itemised terrain, peppered with verifiable and disjointed 'facts' or bits of information. It is the domain of the ancestors, a resonant source for renewing awareness of present existence, which itself is not an itemized terrain either. Orality knows no lists or charts or figures.

But Ong's account is overdone. The predominantly oral culture of twelfth- and thirteenth-century England was by no means unsophisticated in the arts of remembrance and calculation. The oral tradition had ensured the survival of King lists from Anglo-Saxon times. The minor courts and folk-moots all functioned with unwritten precedents handed down from generation to generation. Certain monetary transactions (such as payments made and received) could be recorded with tally sticks. Dates could be marked out via 'clog almanacs'.[3] Further up the social scale, the multiple arts of memory were cultivated (Yates, 1966).[4]

Thus, many of the earliest texts described customs and practices *already in position*. These customs and practices were not immediately transformed by the fact of being written down. The process was gradual. But as the technology of reading, writing and producing texts spread so the consciousness of time past could be extended (and manipulated) and new ways of organising time present (which enabled time to be equated with monetary values) became embedded in everyday practices.

In the twelfth and thirteenth centuries texts can be divided into three types (Clanchy, 1979), as long as it is realised that the delineation of each type is catholic in the extreme. First, there are books; second, 'letters'; and third, 'records'. What seems clear, taking care to avoid equating the currency of these texts with their survival in the historical record, is that all three kinds increased in number over the period and by its end use of certain of these texts had even percolated down to village level.

Books remained rare throughout the twelfth and thirteenth

centuries which, given the manuscript form of production, is no surprise. Perhaps most books were to be found in the sacred realm. Thus Durham Cathedral had 490 volumes in the twelfth century; Christ Church, Canterbury, had 1300 volumes over the same period; Rochester Cathedral had 241 volumes in 1202. It seems likely that some of the monasteries had even larger libraries, especially those of the Dominican order. In the secular realm, King Henry I had a limited library, arguably as large as that of a monastery.[5] But many other kings did without the benefit of books (Clanchy, 1979). A few individuals also possessed books, mainly the nobility and professionals.

> Robert Grosseteste, the greatest English scholar and bishop of the thirteenth century, perhaps had about 90; Master John of Foxton, who sold his books on Scripture to Lincoln Cathedral in c. 1235 possessed about 34 volumes on that subject; Master Peter of Peckham, a lawyer, had 18 works in a coffer when he died in 1293; Guy de Beauchamp, Earl of Warwick, gave away about 40 books in 1306. (Clanchy, 1979, p. 82)

The books in institutional and personal libraries would have covered a range of subjects but they were primarily divided into works of reference, religious books and tables, and Latin classics. Each of these kinds of book promoted a particular attitude to time. Thus, religious works and the classics provided at least some notion of long periods of historical time, although their function as exemplars of events in the Bible and antiquity clouded this perception. But the period also saw the production of books that were specifically concerned with the marking of time becoming more common. A few almanacs circulated (Thomas, 1971; Capp, 1979). These contained information on the days of the week, the months, the fixed Church festivals, movable feasts, astronomical events and astrological changes. There were also the *horologia*, tables giving the length of night and day for every day in the year (Landes, 1983) and in some cases divisions by hours and even smaller units.

The next type of text consisted of letters. Letters can be found in Anglo-Saxon times between Kings. By the twelfth and thirteenth centuries their use had become much broader. Letters were mainly used to record financial transactions, such as charters recording titles to property, chirographs, usually noting money debts, and *brevia* and writs, usually emanating from the Court (which by Henry I's time was issuing 4500 letters a year). Letters as we now understand them probably only became (relatively) common in the thirteenth century, as people became familiar with the new medium. Until that time the spoken word had sufficed.

Systematically compiled records, archives and lists, the third type of text, are a comparatively late arrival on the medieval scene. The exceptions to this observation are the monastic chronicles and cartularies and the records of the King's court. The chronicle, which 'had its origins in the Benedictine preoccupation with the careful regulation of time' (Clanchy, 1979, p. 78), was a historical record of a monastic house. In c. 1188 Gervase of Canterbury described its function:

> The chronicler computes years *Anno Domini* and months and Kalends and briefly describes the actions of kings and princes which occurred at those times; he also communicates events, portents and wonders (cited in Clanchy, 1979, p. 78).

The cartulary recorded the title deeds of the monastery in a book for convenience and as a security against fire. The practice spread from the monasteries to other institutions in England during the twelfth century. At the King's Court (where the chief justiciar and Chancellor was as often as not a notable Cleric) financial records can be found through the twelfth century but cumulative non-financial records (especially legal records) date mainly from the 1193–1205 period for most of which time the innovative Archbishop Hubert Walter was Chancellor, the person who also created central government archives. It seems, then, that the practice of both financial and non-financial record-making and keeping, with all its implications for the consciousness of the past and a calculation of the value of time present, spread out from the Court and so onwards to the bishops and other magnates.

Thus, financial accounts may now seem the most obvious way of stating time as money but outside the Exchequer there was no conception of the keeping of a continuous series of financial records to guide the management of an estate throughout most of the twelfth century. If, for example, a landlord wanted to summarise the circumstances of this estate he undertook a *descriptio* or *kalendarium*, a general description of its workings. However, by the end of the twelfth century such a description could be quite astute, showing all the signs of a management very near to the stage of record-keeping.

Then in the thirteenth century a whole series of milestones are passed. In 1208 the earliest extant accounts of a bishopric are found. The earliest records calculating the profits of actual manors date from the Christ Church, Canterbury, accounts of 1224–5 but such records remain rare until the latter half of the thirteenth century. It is during this same period that financial accounts can be found spreading to the great religious houses (from the 1260s), the Oxford colleges (from 1277) and the towns (from 1256). And 'by the middle of the thirteenth century many magnates' households, both clerical and lay, were keeping daily accounts of expenditure in writing on parchment' (Clanchy, 1979, p. 72).

Non-financial records follow a similar pattern of diffusion to financial records. Legal records are the most important. Custumnals detailing common laws and rules are found early on in the medieval period but the writing down and storing of the details of individual cases does not come in until the late twelfth century. The practice spreads from the King's court to the ecclesiastical courts to the county courts and so finally to the local courts; the general impression left by the evidence 'is that the habit of making and keeping records of proceedings in continuous series stemmed from the King's Court at the end of the twelfth century and took another century to spread across the country' (Clanchy, 1979, p. 77).[6] Lists appear in the same sequence. By the end of the thirteenth century listing had become a commonplace activity. For example, lists of men over the age of 12 were compiled for the view of frankpledge by seignorial bailiffs and stewards and bailiffs were expected to make lists of everything on the manor.

These developments (and many others) show time consciousness developing over the course of the twelfth and thirteenth centuries in the matter of texts. In particular, a sense of time as ordered and as stretching into the past becomes more obvious as 'the business of government, whether ecclesiastical or secular, . . . gradually becomes associated with both learning and written precedents' (Clanchy, 1979, p. 132). Nowhere is this clearer than in the practice of dating documents. Thus, most twelfth century charters do not record dates. If they do, then these dates are clumsily put and related to significant events. It is not until the last decade of the twelfth century that royal letters are uniformly dated (by place of issue, day of the month and regional year). From the court the practice spread to the bishops but it is not until well into the thirteenth century that the practice of dating letters and other texts (such as books) becomes routine (although remaining far from consistent in form).

> Why not always give the date on a document in a precise and uniform way, if only as a routine precaution? Various explanations can be suggested for not doing so. One sometimes mentioned is that charters were seen as mere confirmation of transactions which had already taken place, and there was no point in dating them . . . Another possible explanation is that they were difficult to compute without printed diaries . . . Other reasons for not putting dates on documents are more profound. Until putting the date on a document became a mindless routine, dating required the scribe to express an opinion about his place in time . . . . Paradoxically monks may have so rarely dated documents in the twelfth century or earlier because they were too conscious of the significance of time and of their place in posterity.
>
> The non-religious, by contrast, had such a personal and short view of time that they too found it difficult to specify a numerical year on a document. Everyone knew which year was meant, the present one, and if there were doubt, some notable event could be referred to (Clanchy, 1979, p. 238).[7]

It is tempting to look for some symbolic final seal on the

matter of dating texts. The year 1290 suggests itself, when legal memory was set at the accession of Richard I in 1189, where it was to remain throughout the rest of the Middle Ages. (Up until the reign of Edward I the assumption of the legal system seems to have been that time extended only to the memory of the oldest living person. From 1290 time was set fast in the artificial memory of royal bureaucratic records and archives.) But history is never so neat. The creation of legal memory was a fiction designed to suit royal bureaucrats and only royal courts functioned in that way. Manorial and other folk courts continued to operate with a mixture of written and unwritten precedents which were often little more than a codification of oral memory for some time to come.

In all this, little mention has been made of the towns and the developing merchant class within them. The period from 1100 to 1300 is certainly a period of the growth of industry (although much of this industry was located in rural areas) and of the growth of trade (signified by, for example, the increase in the number of market towns). But, perhaps surprisingly, the growth of industry and commerce in the towns did not lead to rapid increases in the use of documents. The management of boroughs led to the need for documentation, for example summarising municipal bye-laws. The guilds, many formed in the twelfth century, needed lists of guildsmen and summaries of rules. But the practice of trade was still often at a non-textual stage. Contracts were often oral, especially in the twelfth century. Not until 1285 with the Statute of Merchants was every important town in England obliged to have a clerk to enrol debts (although written contracts were certainly quite extensively used before this time). The great merchant dynasties (most of which were knightly) were as educated as the knights over the period and no doubt used documents towards their end but the lesser merchants (who made up the bulk of trade) were a different prospect altogether:

> it is doubtful whether literacy in Latin was yet an essential skill, as they worked from memory and tally sticks. Book learning and book keeping became crucial to lesser merchants only when they ceased to travel with their

> wares and sat in offices instead. On the whole, that is a development of the fourteenth century rather than the twelfth, as far as England is concerned. (Clanchy, 1979, p. 188)

Beginning from the Court and the Church in the twelfth century, the rising tide of documents washes up on the shores of local courts by the end of the thirteenth century, bringing with it new forms of awareness of time. Further, there is evidence to suggest that those who could read and therefore be influenced by the time sense purveyed by the documents were, by the end of thirteenth century, a fairly sizeable group (Thompson, 1963). The extent of medieval English literacy is a notoriously difficult area of research, complicated by such factors as the existence of three languages, the differences between reading and writing literacy, and so on. But, of familiarity with documents as a prerogative of a large proportion of the population there is, by the end of the period, no doubt. One index of this familiarity is ownership of a seal (to authenticate documents). On this index whereas in 1100 only rulers and bishops possessed seals by 1300 all freemen and even some serfs owned them (Clanchy, 1979). Of minimal literacy (comprising reading literacy in a little Latin, and perhaps French), a similar distribution seems at least possible.

In summary, the historical geography of the new forms of time consciousness engendered by texts in the twelfth and thirteenth centuries is the geography, above all, of the presence or absence of particular social ranks. In the twelfth century only the Court and the monastic orders are strongly under the spell of reading and writing. Then gradually through the course of the thirteenth century, the spell's sphere of influence gradually reaches out to touch the bishops and the clerics, then the lay barons, then the gentry and knights and finally the humble peasant, even down to some serfs. The progress of the spell down through the social ranks is uneven, according to the type of document in use, but it is remorseless.

2.2 *Devices*

In the period from 1100 to 1300 there was not only a change in time consciousness emanating from the increase in textual usage associated with the new ideology of everyday practice. As or more important was the increase in devices that marked out the time. Of these devices bells and associated timekeeping devices are by far the most important (Morris, 1955; Downman, 1898; Price, 1983).

Bells were the major markers of time through the period under discussion, nearly all located in churches. 'It was mainly by the ear that medieval people knew the time, not by eye' (Gurevich, 1985, p. 106). The growth in the power of bells as markers of time can be noted in three ways.[8] First of all, there was the sheer number and size of some bells in the chief locations. Thus at Exeter Cathedral by 1050 there were seven bells, to which Bishop Leofric added six more in the twelfth century (Scott, no date). By the twelfth century Canterbury Cathedral was reported to have five large bells which reputedly needed 10, 10, 11, 8 and 24 men to ring them, respectively. Later a larger bell was added which required the services of no less than 32 men![9] In the same century enormous bells could be found at the cathedrals at Winchester, Lincoln, York and St Albans, as well as at the Abbeys of Westminster and Bury St Edmunds (the latter probably destroyed in 1210 when the Abbey towers fell in). Second, the number and size of bells to be found can be indexed through the diversity of church architecture in the period (Rodwell, 1981; Morris, 1955). Bells were housed in three ways. There was the bell-tower, already common in late Anglo-Saxon and early Norman England. Then there was the separate bell-tower (campanile) able to take large bells without threatening the structure of the main building (and found even in major cathedrals like Salisbury and Chester). There was also the bell-turret (bell-cote), often at first made of wood. This construction was ideal for smaller bells. A final index of the number and size of bells in the twelfth and thirteenth centuries is the presence of bellfoundries, found first in England in the tenth and eleventh centuries. Many bells were made by itinerant bellfounders on site. Other bells were made by monks. There is evidence of their activity at Bury St Edmunds, York and Canterbury. Still other bells were made at professional bellfoundries and then transported, possibly along the waterways, to their destinations. The distances involved could be great. For example, bells made at Gloucester ended up in Northampton. The major professional foundries were in London, where the founders congregated in the district between St Andrew Undershaft and St Botolph Aldgate.[10]

In the period from 1100 to 1300 the main source of time marking was therefore the church bell. In the humble life of a small rural community this might only be one bell hung over the roof of the choir. But larger churches would, as has been noted, have many bells. The bells were used to decorate and intersperse services but they also had distinct functions. The most important function of the church bell was undoubtedly to call the celebration of Mass. The service had been held daily from the sixth century in England and bells called for it at noon or afternoon on workdays and on the mornings of Sundays and holy days. At cathedrals and other important churches the bells also summoned people to early morning matins and later some bells were being used to ring out the full set of canonical hours (see below). Thus:

> As cities grew additional masses were celebrated, particularly on Sundays, to service the increased population of the parish. The ringing for these, along with a growing number of religious activities in the church – special sermons, rosaries, novenas, catechisms, public instructions to adults and children – so increased the sounding of bells that in time the air over the housetops was seldom void of the sound of a bell calling to some church. (Price, 1983, p. 107)

To the basic religious ringing were added sets of secular signals. In many places these could be very mundane, for example to indicate that the parish oven was ready. But the most important secular signal was undoubtedly the curfew bell which indicated that all fires must be out and all lights extinguished at a certain time of night. Curfew was imposed by William I and, except for a brief repeal by Henry I initiated in 1100, it was rung throughout the twelfth and thirteenth centuries, except on festival days. The curfew bell sounded at different times in different places. At St Albans in the thirteenth century it was rung at 9 o'clock. This was the hour in London as well and in most large towns. But in the countryside the hour was more flexible (partly because of the lack of time devices). In summer it was at sundown, in winter at a convenient time after dark (Price, 1983).

The dividing line between secular and sacred was a thin one, however. For example the curfew bell gradually became associated with the saying of three 'Ave Marias', a practice formalised in a decree issued by Pope Gregory IX in 1230. (Gradually, however, this practice became separated from curfew and later became the Angelus, with its own bell, with the frequency of the observances increased to three times a day.)

If in the churches the ringing of bells was an important part of life, in the monasteries it was an obsession. In the monasteries the day was, in principle, divided into eight 'canonical hours':

> An office every three hours except for one at midnight made seven offices of 'hours' every day. Some monasteries also held one at midnight . . .. The twenty four hour day started at 6 a.m. . . . or at sunrise, at which the office known as prima (*hora prima*) was held. Three hours later (9 a.m.) came the office of tierce (*hora tertia*); at noon, sext (*hora sexta*); at 3 p.m., none (*hora nona*); and at 6 p.m., or at sunset, vespers (*hora vespera*). Around 9 p.m. compline (*hora completarum*) completed the day before retiring (in some monasteries, before supper and retiring). Where an office was celebrated at midnight it was known as nocturns (*hora nocturna*) or matins (*hora matutina*). (Price, 1983, p. 116)

However, the canonical hours were only very rarely equal in length because the exact times of day at which offices were held often varied. They varied because certain orders were stricter than others. More particularly they varied because the daily round of the monastery was, in the 1100 to 1300 period, still often regulated by hours of daylight. The canonical hours therefore changed with the season (unless a *horologia* was available). For example, the Benedictines rose at 1.30 a.m. and retired at 8.30 p.m. in summer and at 2.30 a.m. and 6.30 p.m. in winter (Dickinson, 1961). Thus there was tremendous variation in the times at which bells were rung:

> Some of the divine offices were taken into the secular churches, and particularly into cathedrals, schools and hospitals. To hold them at convenient times in these places, two or three would be grouped after a Mass, and others shifted in time or omitted. Not so the monasteries, which held them all at fairly regular intervals. Some orders even held additional prayers for laymen and women working in their institutions, known as 'little offices'. Thanks to the monastery bell, the devout within hearing could also share the hours where they were. (Price, 1983, pp. 116–17)

Daily life in the twelfth and thirteenth centuries was therefore conducted by and to the sound of church bells. Le Goff (1980) and others have talked of a change, in the period from 1100 to 1300, in the hours at least, from the time of the church to the time of the merchant, from the sacred to the secular. But it is important not to overstate this tendency. To begin with, medieval English society had no clear idea of a distinction between sacred and secular. Ecclesiastical and lay interests were often integrated. Further, as the number of church bells increased in what were larger and larger towns, surrounding themselves with walls, the attachment to the church bells became, if anything, greater since they were now needed to regulate the new internal threats of fire and disturbance of the peace.

> It was therefore necessary to pass civil enactments in relation to certain church-bell signals, and as these internal changes were greatest at night, the first concerned curfew. We see the religious and secular authorities working together to carry out a law passed by the Common Council of London in 1282 which was that 'at each parish church curfew shall be rung at the same hour

> as St. Martin's (Le Grand) beginning and ending at the same time, and then all the gates, as well as taverns, whether of wine or ale, shall be closed and no one shall walk the streets or places'. (Price, 1983, p. 135)

In certain places, however, secular signals were given via special bells belonging to the town but installed in the church, or even installed in a separate tower set off from the main church building. In addition, by the thirteenth century, bells were starting to be installed outside churches to signal time. They could be found in institutions with fixed timetables like hospitals, on watch and gate towers (where they were rung to signal the opening and closing of the town gates) and on civic and mercantile buildings. Thus, some guild halls had bells. Some of the bells on these buildings were rung at different times from the church bells; for example, to signal transactions. In addition, markets, which had been opened and closed via church bells, were gradually switched to civic bells, although the time when market halls were built with bells atop them is still in the future.

To summarise:

> as commerce developed and industry expanded, the complexity of life and work required an ever larger array of time signals. These were given, as in the monasteries, by bells; the urban commune in this sense was the heir and imitator of the religious community. Bells sounded for start of work, meal breaks, end of work, closing of gates, start of market, close of market, assembly emergencies, council meetings, end of drink service, time for street cleaning, curfew, and so on through an extraordinary variety of special peals in individual towns and cities (Landes, 1983, p. 72).

It is important to stress that in medieval times these bells were not just time signals. There was a moral-psychological dimension to their existence. For example, bells were thought to be able to drive away evil and disease, and able to help boost the prayers of the faithful towards heaven. Thus, bells were often baptised, named and given personal attributes:

> in the Middle Ages the church tower bell was not held to be mere metal. With its reception into Christianity through baptism, with its possession of a name, with its mantle of iconography and holy symbols exposed to the powers of the air, with its voice which spoke to heaven and earth in a distinctive tone known to the whole community (for no bell sounded quite like any other), with the translation of this sound into words on its surface, and above all with the trust that the whole community had in the effectiveness of these attributes, the bell was regarded as a half-divine being with a personality. It was therefore natural that any inscribed statements of the bell's purpose or power which were cast on it when it was made should be made in the first person.
>
> These statements give us a glimpse of life in former times from a little known vantage point. They were called virtues, literally (strengths), but they ranged from the common functions of a bell to stupendous miracles . . . . It did not matter if they were in a speech not understood by (some of) the people. They were not there to be viewed by men, but to be spoken to angels and demons contending for men's souls. (Price, 1983, p. 127)

Bells had to be rung at particular times but during the period from 1100 to 1300 these times were difficult to determine. Thus a London-based common law judge, Roger of Seaton, acknowledged receipt of one letter from the Chancery in 1279 at 'about the ninth hour' and another 'just a little after dark' (Clanchy, 1979, p. 236). The passage of the day – sunrise, noon and sunset (in clear weather) – was still, bells or no, a powerful timekeeper. But other timekeepers existed to help keep the bells accurate. The sundial was ubiquitous and was used with full knowledge that days became longer as the year approached the solstice and shorter as it left; on some sundials lines pointed out the differences. A very few persons had access to *horologia* and even astrolabes (White, 1962). Of course, days were not always clear and sundials were of no use at night, so there was also the hourglass and the candle marked with equidistant lines. Both of these devices had to be carefully watched and it was even possible to fit primitive alarm mechanisms to them to keep the watcher alert. This was especially important in monasteries with night offices to be celebrated. In the monasteries, prayers chanted at a fixed speed and rhythm covered a fixed period of time and were a useful supplementary timing device.

Nearer to the apex of sophistication was the water clock (clepsydra). There was one such clock at the Abbey at Bury St Edmunds in 1198, for example. It is recorded because its water helped to extinguish a fire but there were certainly others at the same time, records for which have not survived, especially in cathedrals and monasteries. By the thirteenth century these water clocks were of a considerable sophistication (White, 1962). Finally, right at the end of the thirteenth century, mechanised clocks begin to appear. These early clocks could be major engineering projects so it is no surprise that records of their construction or other mention of their existence has survived. Beeson (1971) argues that the earliest known mechanical clock in England was located at Dunstable Priory. It is first mentioned in records in 1283. Other such clocks are first mentioned at Exeter Cathedral (1284); Old St Pauls, London (1286); Merton College, Oxford (1288?); Norwich Cathedral Priory (1290); Ely Abbey, a Benedictine house, in 1291; and Christ Church Cathedral, Canterbury in 1292 (North, 1975). Mechanical clocks then appear regularly through the fourteenth century, starting out at Salisbury Cathedral (1306), Norwich Cathedral (1322), Lincoln Cathedral (1324) and St Albans Abbey (1327) (Figure 1). At this early stage, most of the clocks proper had no dials and only counted hours. Further, they were not always linked directly to outside bells. Sometimes, they acted as alarms to get the bell ringer to the bell (Beeson, 1971; North, 1975). But, by the early fourteenth century directly linked clocks had appeared, at least to judge from a 1306 regulation from Sarum (Salisbury) which states that 'before the clock of the Cathedral had struck one no person was to purchase or cause to be purchased flesh, fish or victuals' (Beeson, 1971, p. 16). The new mechanical clocks also announced the prestige of their owners in no uncertain manner. As Landes (1983)

Figure 1: Date of first report of mechanical clock

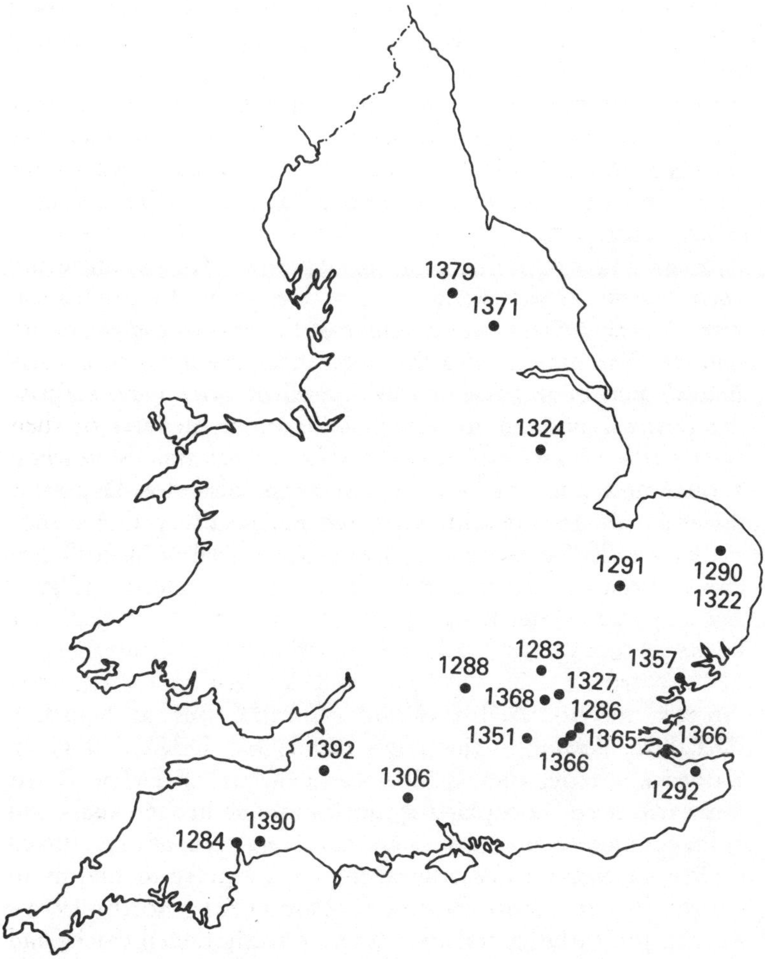

(Source: Beeson, 1971)

points out, this totemic function was sometimes as important as the timekeeping function.

Clearly, the use of water clocks and mechanical clocks from the thirteenth century onwards had important consequences for time consciousness. In particular, it meant that, for the first time, the passage of time could be followed independently of the passage of night and day through an organisation of the day into strictly equal periods of time. Thus, a wholly artificial system of time reckoning was possible for the first time in daily life. This had certain consequences and most particularly, it helped to measure and calculate durations of time and allocate this time in new, more exact ways.

In summary, the historical geography of the new forms of time consciousness engendered by devices in the twelfth and thirteenth centuries is, in the main, the geography of the church. The church and the associated monastic orders (see below) were more likely to have need of bells and timekeeping devices. Because so many of these devices were publicly in earshot or on display, sacred time was also increasingly secular time and nowhere more so than in the towns. But, in general, it was not until the fourteenth century that certain bells and clocks became totally secular devices as bells and clocks began to be found on, for instance, market or guild buildings or at the King's Court.

2.3 *Systems of discipline*

The third element of the newly developing ideology of everyday time practice consists of systems of time discipline. There were two main exemplars of time discipline in the twelfth and thirteenth centuries, one sacred and one secular.

The sacred exemplar was the monastic system with its offices set at canonical hours. As pointed out above, the strictness of the monastic system of time discipline varied from order to order according to the rigour with which each order applied the Rules of Saint Benedict, which had been extant in English monasteries from the seventh century. The monastic attitude to time is summed up in Rule XLVIII of *The Rule of Saint Benedict*, 'Idleness is the enemy of every soul' (McCann, 1970, p. 28).

The timetables (*horaria*) of the monastery run along the lines of the Rule of Saint Benedict were therefore carefully planned out to make sure that the brothers were never idle. First of all, there was an annual timetable.

> There was a separate *horaria* for the summer and for the winter, and even three separate ones with respect to mealtimes[11] . . . there were also various routine activities such as bathing, bloodletting, head shaving, and mattress refilling, which were performed on a regular basis a fixed number of times during the year. (Zerubavel, 1981, p. 33)

Second, there was also a weekly timetable. Certain psalms were said only on certain days of the week.

> Further, mealtimes and wake-up time also varied in accordance with the day of the week, and even feet and beer jugs were to be washed on a weekly basis. To appreciate the weekly 'beat' of monastic life, note also that the reading of the full cycle of 150 psalms began on Sunday, as did the rotations of kitchen service and of reading to the community during meals. (Zerubavel, 1981, p. 34)

Finally, there was the daily round. The offices associated with the canonical hours (for which there were careful rules of sequence) were the framework around which a whole set of activities were spun. The daily order of prayer was only a part of the ordering of all daily life. Indeed for monks there was no distinction between the religious and the worldly – work was not an end in itself, rather *laborare est orare*, to work is to pray. So there were rules setting aside times for work, study, eating, sleeping, as *The Rule of Saint Benedict* makes clear in almost neurotic detail:

> The brethren . . . must be occupied at stated hours in manual labour, and again at other hours in sacred reading. To this end we think that the times for each may be determined in the following manner . . . the brethren shall start work in the morning and from the first hour

> until almost the fourth do the tasks that have to be done. From the fourth hour until about the sixth let them apply themselves to reading. At the sixth hour, having left the table, let them rest on their beds in perfect silence. (McCann, 1970, p. 39)

At night:

> the brethren shall rise at the eighth hour . . . so that their sleep may extend for a moderate space beyond midnight . . . let the hour of rising be so arranged that there may be a short interval after Matins, in which the brethren may go out for the necessities of nature. (McCann, 1970, p. 41)

The rules for lateness were strict with penalties in the form of penances for latecomers (see Zerubavel, 1981):

> Why was punctuality so important? One reason was that lateness – 'God forbid!' – might make it necessary to abridge an office, in particular matins: 'let great care be taken that this shall not happen'. Another, I think, was that simultaneity was thought to enhance the power of prayer. (Landes, 1983, pp. 62–3)

Twelfth- and thirteenth-century England was subject to successive invasions by monastic orders establishing themselves in the wake of the Norman Conquest. By the end of the period the monastic population had reached some 18,000 persons (including nuns) (Dickinson, 1979). Thus after the Benedictines (who were present in England before the Conquest) came a number of other orders whose common feature was their willingness to adhere strictly to, and even build upon, the Rule of Saint Benedict as other orders gained wealth and lost their enthusiasm for rigour. In the Cluniac monasteries:

> the now traditional Benedictine insistence that nothing was to be given precedence to the maintenance of an elaborate round of worship was much amplified by the adoption of an immensely complex music and various expansions in the texts of the offices which led to communal worship completely dwarfing all other aspects

of the Clunian monastic life. (Dickinson, 1979, p. 102)

After the Cluniacs came the Cistercians, an austere order keeping to remote places and following the Rule of Saint Benedict *ad apicem litterae*, to the last letter. The list goes on.[12]

The importance of the system of time discipline in the monasteries spilt over into the estates in their care which were run on the same precise temporal lines. Later in the period lay brethren extended their concern for proper timekeeping even into non-monastic estates (see Capp, 1979). Thus the Cistercians divided their communities into monks (*monachi*) and lay brethren (*conversi*). The lay brethren acted as a bridge over the religious/secular divide, carrying time discipline into the countryside. It also seems likely that the vision of time of orders like the Benedictines, Cluniacs and Cistercians would have made its way into the running of ecclesiastical and royal estates, if nowhere else.

The second major exemplar of time discipline was secular, insofar as the word had meaning in twelfth- and thirteenth-century England, and consisted of a set of interconnected changes in the functioning of the medieval economy. Of these changes by far the most important was the change in the calculation of time economy allowing for the intensification of work practices.

In the countryside, work hours tended to be from sunrise to sunset but the more important event here, so far as work practices were concerned, was the attempt to calculate work hours as units of labour time with a set value. (For more detail, see Coulton, 1926 and 1938; Bennett, 1956; Hallam, 1981; Homans, 1941.) Obviously, before the thirteenth century the fact that the serf (and in some regions the free at set times of the year) had to render labour services meant that measurement of the value of time was needed to some degree, as Coulton (1938, p. 73) makes clear:

> For those unfree there was one general rule . . . that principle was, that the serf possessed nothing of his own; yet he might enjoy his land so long as he rendered three days work in the week to his lord. Before those days of

> the thirteenth century, when the multiplication of records enables us to get a far clearer view of the details, it is evident that a so-called 'day's work' had been reduced to a half-day of real honest labour. At Great Chesterford in 1270, each villein owed 714 week-works, (the main unit of labour services, rendered week by week through the year) each of which occupied a quarter of his day. In addition, at ploughing time he had to plough 16¼ acres. This totals at nearly three days a week. Some of these customary burdens are very complicated indeed; here, for instance, are the requirements on a Glastonbury manor. Between Michaelmas and November 11th the serf had to plough one acre; another half acre when called upon. He must harrow every Monday in the year; three times a year he must carry a load for my lord; he must mow three days, and go to the vineyard. All this he must do anyhow; then comes a rather more complicated list of works for which he may compound, if he chooses, by a payment of 3s. 4d. a year.

In such a system, money value measures of output per unit of time were still comparatively crude (as indeed were the measures themselves; see Coulton, 1926, Chapter 5; Kula, 1986). But such measures became more sophisticated through the thirteenth century for two main reasons. The first was the increasing commutation of labour services for money and especially the payment of money rents. The second was the increasing use of surveys and the keeping of records. By the end of the century calculations were routinely recognising time as money. For example, by 1303 the records of the Cuxham manor demesne note 1323 customary labour service 'dayworks' (a measure of area, probably the amount of land customarily delved in a day) carried out at a value of £7.2.10½ as well as 2060 dayworks for the *famuli* (full-time workers) and payments to unskilled hired labour made partly in cash and partly in corn. Tasks like ditching, muckspreading and threshing were charged at one daywork rate, while tasks like reaping or carrying were charged at another rate (Harvey, 1965).

But the system of time discipline and valuation was

undoubtedly more rigorous in the towns where, by the end of the thirteenth century probably one tenth of the population of England could be found (Treharne, 1971). In the towns work hours were generally accepted in the guild statutes of the time as from dawn until 9 p.m. in winter (Michaelmas to Easter) or from sunrise until 30 minutes before sunset in summer (Bienefeld, 1972) with a break in both cases at midday. But, within these bands, there were great variations according to type of work and the skill of the worker. For example, some workers might work less time, while other workers – if the project was urgent – might work on by artificial light. Work days were usually from Monday to Saturday, with some workers taking Monday off as well, the beginning of the practice of 'Saint Monday' (Reid, 1976). But, of course, there were feasts and saints' days to be taken as well.[13] Some of these days, like Christmas and Easter, were nationally taken, but others were local in character. For example, the dedication saint's day of the local church was usually a local holiday. Thus in 1297 the workmen at Windsor did not work on Monday 7 October which was 'the feast of dedication of the church in the township' (cited in Salzman, 1967, p. 59). There was also a general custom that when two or more feast days coincided the workmen received wages for one day in every two holidays.

Regulations about length and value of work hours became remarkably exact as writing became more common. Not so far outside the period, in 1352 for example, the craftsmen employed at York Minster were enjoined to follow these elaborate regulations, worth quoting in detail:

> That the masons, carpenters and other workmen ought to begin work, on all working days in the summer, from Easter to Michaelmas, at sunrise and ought to work from that time until the ringing of the bell of the Blessed Virgin Mary, and then they should sit down to breakfast in the lodge of the works, if they have not breakfasted, for the space (of time that it takes to walk) half a league; and then the masters, or one of them, shall knock upon the door of the lodge, and all shall at once go to their work; and so they shall diligently carry out their duties

> until noon, and then they shall go to their dinner (*prandia*). Also, in winter, from Michaelmas to Easter, they shall come to their work at dawn and everyone when he comes shall immediately start work, and so continue in the said way until noon. From the feast of the Invention of the Holy Cross [3 May] to the feast of St Peter's Chains [1 August], when they ought to sleep in the lodge after dinner; and when the vicars come out from the canon's hall (*mensa*) the master mason, or his deputy, shall cause them to rise from slumber and get to work; and so they ought to work until the first bell for vespers, and then they shall sit and drink in the lodge, from the said first bell to the third bell, both in summer and winter. Also from the [1 August] to the [3 May], they shall return to their work immediately after their dinner, for which a reasonable time shall be taken, without waiting for the return of the vicars from the canon's hall; and so they shall work until the first bell for vespers and then they shall drink in the lodge until the third bell has rung, and shall return to their work, and so they shall work until the ringing of the bell of St. Mary's Abbey, which is called Le Langebell, namely, every working day from the feast of St. Peter's Chains to Michaelmas, and from Michaelmas to the said feast of St. Peter, they shall continue to work as long as they can see by daylight. Also each mason shall take less for the week in winter, that is from Michaelmas to Easter, than in summer by one day's wage. (Salzman, 1967, pp. 56–7)

Rates of pay in the towns show a general appreciation of the value of labour time. As the passage shows, greater rates of pay were awarded: in summer than for the shorter days of winter time; for different types of skill; for foremen (who all received paid holidays); for overtime (called 'respite hours'); and at harvest time (when many were liable to work in the countryside and had to be attracted back).[14] It was also quite traditional to give gifts (usually of drink, but also of money and clothing) to encourage industry, although it is important to point out that these gifts were no innovation but rather relics of the *corrodium* (food and drink) given to boon workers as subsistence when in the fields on the lord's work.

Other changes in the medieval economy also had repercussions for time consciousness. On the subject of the towns, it is important to remember the impact that the expansion of trade, finance and industry was making upon the boroughs and market towns, especially in the thirteenth century. The expansion of trade (especially in wool) brought with it periodic market systems and trade networks of considerable intricacy and correspondingly more exact and more exacting timetables.[15] For merchants, for example:

> Once commercial networks were organised . . . time became an object of measurement. The duration of a sea voyage or a journey by land from one place to another, the problem of prices which rose or fell in the course of a commercial transaction (the more so as the circuit became increasingly complex, affecting profits), the character of the labour of craftsmen and workers (since the merchant was almost always an employer of labour), all made increasing claims on (the merchant's) attention and became the object of ever more explicit regulation. (Le Goff, 1980, p. 35)

New monetary instruments and forms of credit (overcoming the medieval distrust of usury) were also introduced which required more adequate measurement of time.

The expansion of urban industry was another notable feature of the thirteenth century (although a corresponding increase in rural industry also took place) and the new processes of production also needed exact time measurement. These new demands for more exact time measurement could have reciprocal effects. For example, they could be called into service in measuring the output of workmen and women. Time therefore became thought of independently of what was happening in time. Finally, the expansion of the medieval state over the period from 1100 to 1300 deserves mention, bringing with it new and more complex judicial and other timetables which intermeshed with and regulated those of the economy.

In summary, the historical geography of the new forms of time consciousness born out of the development of systems of discipline had its roots in the exact and exacting regimen of the monasteries and the calculative demands of the towns. From these two 'lighthouse' sites the idea of time discipline makes its way outward into a whole set of different contexts. Most obviously, the monastic idea of time discipline makes its way into associated institutions; the monastic estates, hospitals and so on. But, more indirectly, it also makes its way into the royal bureaucracy (still run by the Church) and, through lay brethren, into other estates. The idea of such discipline would also have lain heavy over the towns, characterised by the ringing of the canonical hours. But, as Foucault (1977, p. 154) points out, the idea of time discipline in the monastery was essentially *negative*: 'it was the principle of non-idleness: it was forbidden to waste time, which was counted by God and paid for by men; the timetable was to eliminate the danger of wasting it – a moral offence and economic dishonesty.' The system of time discipline adopted in the towns (although clearly influenced by the monastic regimen) was rather different in focus. It was becoming orientated towards a more modern, *positive* use of time, based upon the intensification of work 'as if time, in its very fragmentation, were inexhaustible or as if, at least by an ever more detailed internal arrangement, one could tend towards an ideal point at which one maintained maximum speed and efficiency' (Foucault, 1977, p. 154). Clearly, in the twelfth and thirteenth centuries, this process still has a long way to go but certain elements of such a system are already in place; for example the ability to value use of time and, consequently, to begin to measure productivity (see Tribe, 1981). As Gurevich (1985, p. 150) puts it: 'Time was recognised as a commodity of great value and as a source of material gain.'

3 CONCLUSIONS

The period from 1100 to 1300 is a period when the ideology of everyday time practice is changing and with it the consciousness of time. There are several indices of this change. First, there are the practices associated with the writing and reading of texts, as in the cases of the dating of documents and the keeping of records. As a result, for those exposed to texts, the past was able to be extended further backwards while the present became more predictable and more open to manipulation. Second, there are the practices associated with the keeping of time. Timekeeping devices not only became more accurate but inspired a new idea of time as an artificial reckoning grid, encapsulated in the advent of mechanical clocks and equal hours. As a result, the passage of time could be more exactly measured with consequences for the calculation of labour time and the spawning of attempts to make time more productive. Finally, there are the routinised practices associated with new systems of time discipline and valuation. These become more rigorous over the twelfth and thirteenth centuries. As a result, the planning of the position of people and objects over periods of time, an activity that had started with *hidation* and *carucation* in the seventh and eighth centuries become more and more sophisticated. In a sense, time-geography became possible.

These changes were all, of course, inter-related in complex ways. To take one example only, effective systems of time-discipline need some form of timekeeping. The more ambitious these systems become the more a need for exact timekeeping will be felt. Mumford's (1934) classic statement that the invention of the clock is linked to the rigour of the monastic system of time discipline therefore still rings true (see the discussion of this point in Landes, 1983, pp. 53–7). But the relationship between systems of time discipline and timekeeping devices was reciprocal. The invention of time-keeping devices gradually opened up possibilities of new interpretations of time leading to new systems of time discipline. Landes (1983, pp. 77–8) puts this well:

> Medieval man (sic), it has been observed, was innumerate as well as illiterate. How much reckoning could he do in a world that knew no uniformity of measurement? Units of distance were linked to physical characteristics that vary as people do (the English *foot*, for instance, and the French inch, called a *pouce*, which means thumb); while weights typically were converted to volume standards (a *bushel* of grain) that inevitably varied from place to place and mill to mill. Even the learned were not accustomed

to using numbers. The calculation of the calendar, for example – a crucial aspect of liturgical discipline – was confined to specialist computists. The schools offered little if any training in arithmetic, and persistence of roman numerals was both symptom and cause of calculational paralysis.

All of this began to change in the twelfth and thirteenth centuries . . . .. This was a period of growing trade and he who trades must reckon. So must clerks and functionaries who count taxes and expenditures, and these were years of rapid development of royal power and government apparatus.

It was the urban, commercial population that seems to have been quickest to learn the new language and teachings. Arithmetic was the province above all of the intellectual speakers of the vernacular (as opposed to Latin). Many of these learned arithmetic in the shop or on the road, but even before they entered trade, they learned to count by the bells or the clock. Not by the old church bells ringing the canonical hours; these did not mark equal units and hence did not lend themselves to addition or subtraction. But the new bells and the calculations they made possible (how long until? how long since?) were a school for all who listened and began to organise their lives around them. Meanwhile the church clung to old ways and, so doing, yielded the rhythm of life and work to the lay authorities and the bourgeoisie. Equal hours announced the victory of a new cultural and economic order. Here indeed was an unintended consequence: the monks had wrought too well.

Above all, the new time consciousness wrought by the complex interplay between new expressions and new mediums of expression was formulated as accounts, accounts pregnant with new possibilities (how long until? how long since?), accounts containing new metaphors (time as a clock, time as money), accounts presaging greater use of past and future oriented tenses. Such evidence as we have of these changes in interpretation is largely indirect; we cannot capture shifts in the conversations of the past easily at this distance in time from them. More than now, 'the written record does not give us the whole of medieval man's (sic) way of perceiving and representing his world which has been irretrievably lost' (Gurevich, 1985, p. 37). Thus documents can only give, at least, tantalising glimpses. For example:

> the drafting rule became general that the past tense should be used in charters for the act of giving: 'know that I, A of B, *have* given', not simply 'I give'. This emphasised that the ceremonial conveyance was the crucial transaction, whereas the charter was merely a subsequent confirmation of it. This rule only became general in the thirteenth century. Numerous charters of the twelfth century depart from it, presumably because their amateur draftsmen did not appreciate the relationship between written record and the passage of time (Clanchy, 1979, p. 208).[16]

As has been shown the new forms of time consciousness seen, in this paper, as new accounts of time have a geography, or rather a set of inter-connected geographies. First of all, there is a geography of time consciousness born of social rank. New accounts of time were more likely to be exact amongst the court and the church. The lay nobility, the knights and the merchants were next in line. Finally there was the peasantry. Where these different groups are found the new time consciousness is more or less likely to flourish. Second, there is a geography of time consciousness born of the church. The church and the monasteries had a need for precise timekeeping which often spilt over into the secular medium. Finally, there is a geography born of the growth of the towns. In the period from 1100 to 1300 the urban population of England and Wales grew significantly (if only in line with general population growth). By 1300 London had a population of 50,000; Bristol had a population of 17,000, and York's population was 8,000. By 1300, again, the number of boroughs had increased to 480 from 111 in 1086. Most of these were rural market centres but in the few centres of population like London or Bristol large enough to be relatively insulated from rural rhythms, the new accounts of time took root amongst the populace.

It would not do to exaggerate these points. By the end of the thirteenth century, most of England was still rural and still in train to the rhythms of sowing and harvesting and the bells tolling out the time of Mass. These natural and sacred rhythms still cast a powerful spell over the English social formation. Nor should it be thought that Le Goff's (1980) 'time of the merchant' had yet taken over from 'the time of the church'. At this stage in English history it was more the case that the time of the church was being taken over by the merchant. The schism between sacred and secular time was still some way off.[17] The sacred calendars, the bells calling to mass, the eschatology of Judgement Day; these still exerted a powerful hold on the medieval imagination.

Two final points are in order, one procedural and one substantive. As a first point, time consciousness can clearly never be seen as a unitary phenomenon. Different accounts of time develop at different times amongst people according to the vagaries of history, membership of particular social groups and geography. In the twelfth and thirteenth century certain texts containing certain interpretations of time reach out only to certain people and places; certain devices are the preserve of only certain people and places; certain systems of discipline are only to be found touching certain social groups and places. The accounts people can give of 'time' must depend on the practices they can be involved with in a particular context and the way these practices are influenced by a peculiar mix of local social relations.

Second, some doubt must be cast up on the standard Marxist account of the genesis of modern time discipline and consciousness in England, as first formulated by E.P. Thompson in 1967, with its emphasis on industrialisation as the pivotal event (see Thrift, 1981). For what the account of the medieval ideology of time practice laid out above shows is that significant portions (if by no means the majority) of the population of England were already changing their accounts of time towards quite modern interpretations quite early on in history. Texts were becoming more time-aware. Timekeeping devices were becoming more sophisticated. Rigorous systems of time discipline were already regulating work habits, and not only in the towns; think only of the example of the Benedictine and Cistercian estates. Time was valued, in however primitive a fashion. There is even evidence of conflict over the ownership of time – why otherwise the need for work hour regulation? Harrison (1986) has already come to similar conclusions for the eighteenth and early nineteenth centuries. He notes, for example, that 'work habits were, in fact, far from irregular and haphazard' (Harrison, 1986, p. 174) as the Thompson account implies. Similarly, Smith (1986) provides evidence from pre-industrial Japan of sophisticated attitudes to time at variance with Thompson's thesis. It becomes possible to suggest that Thompson made two major errors in his 1967 paper. First, he underestimated the rigour of pre-industrial work patterns. Even in the medieval period, the evidence is that work patterns were rarely sporadic or wilful, given the need to conform to the pattern of the seasons. Second, he conflated lack of ability to measure time precisely with lack of perception that time can be measured precisely. The evidence from the medieval period suggests that many people were quite aware that time could be measured more accurately. They were simply unable to do so. Thus the Thompson thesis that modern (or in Foucault's term, positive) time discipline finally appeared in its full form with capitalist industrialisation is questionable, to say the least.

Harrison (1986) has suggested that it is urbanisation and, in particular, the strictly timed contexts of the larger towns, that holds the key to the advent of modern time discipline, rather than industrialisation and, on the available evidence, this conclusion certainly seems more realistic. As this chapter has shown, it can be extended back, in embryo form at least, even to the towns of the medieval period. It is in these towns that the institutions of government, church, trade and industry were most likely to come together, increasingly acting to impose artificial schedules on a population which then began to talk about these schedules as natural occurrence.

ACKNOWLEDGEMENTS

David Austins's knowledge and understanding of the medieval period has been an invaluable source in the writing of this paper. Ron Johnston led me to many interesting works on bells. As usual, Derek Gregory made many interesting and valuable suggestions. Participants in a seminar at University College London helped me to hone my arguments. The usual disclaimers apply.

NOTES

1 'I call the living.' One of the most popular inscriptions on medieval church bells.
2 The idea of accounting is complex, as are its antecedents. See Thrift (1986) for a more detailed exposition.
3 Clog almanacs were carved sticks or rods on which notches and symbols provided a calendar showing the lunar cycle and the Christian feasts. Some were small enough to fit in the pocket. Most were larger versions which hung on the wall (Capp, 1979).
4 The production of texts had definite impacts on these acts of memory, a case of the power of oral devices being enhanced by writing.
5 All estimates of numbers are subject to qualification. In particular what was called a 'book' could vary radically from a bound volume through an unbound quire to a table containing but a few pages.
6 Even so, proclamations of laws by the town crier would continue for some time yet.
7 To these reasons might be added others. First, writing only slowly became a trusted medium. Second, exactitude of temporal reckoning was not necessary as long as the temporal infrastructure of documents stayed loose. It was only with the advent of exact administration that exact reckoning became necessary. It is important to remember that by the end of the thirteenth century, births were still not recorded and if age of a person had to be proved in courts (for example, in the matter of wardships) then the normal practice was to ask acquaintances to date births by significant events.
8 A major problem in considering the spread of bells is that, at this time, significantly, bells were rarely dated. The earliest dated bell in England is for the year 1255 (Downman, 1898).

9 These bells were worked by treadle planks. Hence the number of men needed to ring them.
10 Billiter Street in EC3 remains to remind us of their location, a corruption of 'bellytere'.
11 *The Rule of St Benedict* (McCann, 1970, p.23) specifies:

> From the feast of Easter until Pentecost let the brethren dine at the sixth hour and sup in the evening. From Pentecost throughout the summer . . . let them fast on Wednesdays and Fridays until the ninth hour; on the other days let them dine at the sixth hour.

A major complicating factor in all this was start of the migration of *hora nona* during the period to become, midday. Le Goff (1980, pp. 44–5) offers an explanation that the change emanated from the towns, where the pressure for a marker at midday, half the whole day, was great. Alternative explanations by Landes (1983, pp. 404–5) and Gurevich (1985, p. 106) seem just as plausible. This is clearly an area in which debate will be fierce for years to come because there is so little evidence of the reasons for the change.
12 Not forgetting the nunneries, of course. See Dickinson (1979), Rowley (1983).
13 If Sundays, Feast Days, and Saint's Days are added together then up to a third of the year was taken off from work. As Gurevich (1985) points, out, this demonstrates that the work ethic was still subordinate, in many ways, to the demands of religion.
14 Thus the guilds were still as much concerned with quality of work and quantity: 'Work in a guild had not yet degenerated into a total indifference towards its objective content' (Gurevich, 1985, p. 267). Use value and exchange value were not coincident.
15 It is worth noting that the Cistercians, with their rigorous notions of time discipline, were important agents in the wool trade.
16 A glance at the *Oxford English Dictionary* seems to provide some evidence that the vocabulary of time changes quite markedly from the thirteenth century onwards. 'Clock' (from cloche, French for bell) begins to take on its modern meaning. Particular phrases start to circulate such as 'times to come' (first found in 1340) and 'what is the time?' (1200). But, unfortunately though understandably, these dates only record the first written occurrence of a particular word or

phrase. Here the whole problem that new classes of documentation could record older words or phrases for the first time becomes a hideous complication.

17 The symbolic turning point is probably the dissolution of the monasteries.

REFERENCES

Attali, J. (1982), *Histoires du Temps*, Paris, Fayard.

Bagley, J.J. (1960), *Life in Medieval England*, London, Batsford.

Barrows, C. (1972), 'Geography as human ecology', *Annals of the Association of American Geographers*, vol. 13, pp. 68–83.

Beeson, C.F.C. (1971), *English Church Clocks, 1280–1850 History and Classification*, London, Antiquarian Horological Society.

Bennett, H.S. (1956), *Life on the English Manor. A Study of Peasant Conditions 1150–1400*, Cambridge, Cambridge University Press.

Bienefeld, M.A. (1972), *Working Hours in British Industry. An Economic History*, London, Weidenfeld & Nicolson.

Capp, B. (1979), *Astrology and the Popular Press*, London, Faber.

Clanchy, M.T. (1979), *From Memory to Written Record. England 1066–1307*, London, Arnold.

Colson F.H. (1926), *The Week*, Cambridge, Cambridge University Press.

Coulton, G.G. (1926), *The Medieval Village*, Cambridge, Cambridge University Press.

Coulton G.G. (1938), *Medieval Panorama*, Cambridge, Cambridge University Press.

Dickinson, J.C. (1961), *Monastic Life in Medieval England*, London, Adam & Charles Black.

Dickinson, J.C. (1979), *An Ecclesiastical History of England. The Later Middle Ages from the Norman Conquest to the Reformation*, London, Adam & Charles Black.

Dodgshon, R.A. (1978), 'The early middle ages, 1066–1350', in R.A. Dodgshon and R.A. Butlin (eds), *An Historical Geography of England and Wales*, London, Academic Press, pp. 81–117.

Dove, R.H. (1982), *A Bellringer's Guide to the Church Bells of Britain*, (sixth edition), Aldershot, Viggers.

Downman, E.A. (1898), *Ancient Church Bells in England*, Issued privately by author.

Ellacombe, H.T. (1975), *The Church Bells of Somerset*, Exeter.
Evans-Pritchard, E.E. (1940), *The Nuer. A Description of the Modes of Livelihood and Political Institutions of a Nilotic People*, London, Oxford University Press.
Foucault, M, (1970), *The Order of Things*, London, Tavistock.
Foucault, M. (1977), *Discipline and Punish. The Birth of the Prison*, London, Allen Lane.
Giddens, A. (1979), *Central Problems in Social Theory. Action, Structure and Contradiction in Social Analysis*, London, Macmillan.
Giddens, A. (1984), *The Constitution of Society. Outline of the Theory of Structuration*, Cambridge, Polity Press.
Glasser, R. (1972), *Time in French Life and Thought*, Manchester, Manchester University Press.
Gregory, D. (1985), 'Suspended animation: the stasis of diffusion theory', in D. Gregory, and J. Urry, (eds), *Social Relations and Spatial Structures*, London, Macmillan, pp. 296–336.
Gurevich, A.J. (1985), *Categories of Medieval Culture*, London, Routledge & Kegan Paul.
Hagerstrand, T. (1974), 'Commentary' in A. Buttimer, *Values in Geography*, Washington, D.C., Association of American Geographers.
Hagerstrand, T. (1982), 'Diorama, path and project', *Tijdschrift voor Economische en Sociale Geografie*, vol. 73, pp. 323–39.
Hall, D. (1982), *Medieval Fields*, Shire Publications, Princes Risborough.
Hallam, H.E. (1981), *Rural England 1066–1348*, London, Fontana.
Harré, R. (1984), 'Social elements as mind', *British Journal of Medical Psychology*, vol. 57, pp. 127–35.
Harrison, M. (1986), 'The ordering of the urban environment: time, work and the occurrence of crowds 1790–1835', *Past and Present*, no. 110, pp. 134–68.
Harvey, D. (1985), 'Money, time, space and the city', in *Consciousness and the Urban Experience*, Oxford, Blackwell, 1–35.
Harvey, P.D.A. (1965), *A Medieval Oxfordshire Village. Cuxham 1240–1400*, London, Oxford University Press.
Hawley, A.H. (1950), *Human Ecology, A Theory of Community Structure*, New York, Ronald Press.
Hearn, F. (1978), *Domination, Legitimation and Resistance. The Incorporation of the Nineteenth Century Working Class*, Westport, Connecticut, Greenwood Press.

Homans, G.C. (1941), *English Villagers of the Thirteenth Century*, Cambridge, Mass., Harvard University Press.
Kern, S. (1983), *The Culture of Time and Space 1880–1918*, Cambridge, Mass., Harvard University Press.
Knowles, D. (1949), *The Monastic Order in England*, Cambridge, Cambridge University Press.
Kula, W. (1986), *Measures and Men*, Princeton, Princeton University Press.
Lakoff, G. and Johnson, M. (1980), *Metaphors We Live By*, Chicago, University of Chicago Press.
Landes, D.S. (1983), *Revolution in Time. Clocks and the Making of the Modern World*, Cambridge, Mass., Belknap Press.
Langdon White, C. and Renner, G.T. (1936), *Geography. An Introduction to Human Ecology*, New York.
Lave, J. (1986), 'The values of quantification', in J. Law (ed.), *Power, Action and Belief. A New Sociology of Knowledge?* London, Routledge & Kegan Paul, pp. 88–111.
Law, J. (1986), 'On the methods of long-distance control: vessels, navigation and the Portuguese route to India', in J. Law (ed.), *Power, Action and Belief. A New Sociology of Knowledge?* London, Routledge & Kegan Paul, pp. 234–63.
Leclercq, J. (1975), 'Experience and interpretation of time in the early Middle Ages', in J. Sommerfeldt, L. Syndergaard, and E.R. Elder (eds), *Studies in Medieval Culture*, vol. 5, Kalamazoo, Medieval Institute, Western Michigan University, pp. 9–19.
Le Goff, J. (1980), *Time, Work and Culture in the Middle Ages*, Chicago, University of Chicago Press.
Lowe, D.M. (1982), *History of Bourgeois Perception*, Chicago, University of Chicago Press.
McCann, J. (trans.) (1970), *The Rule of Saint Benedict*, London, Steed & Ward.
Marx, K. (1973), *Grundisse*, Harmondsworth, Middlesex, Penguin.
Morris, E. (1949/50), 'British bellfounders', *Hinrichsen's Musical Year Book*, vol. 6, London, Hinrichsen, pp. 294–307.
Morris, E. (1955), *Towers and Bells of Britain*, London, Robert Hale.
Mumford, L. (1934), *Technics and Civilization*, New York, Harcourt Brace.
North, J.D. (1975), 'Monasticism and the first mechanical clocks' in J.T. Fraser and N. Lawrence (eds), *The Study of Time* 2,

Berlin, Springer Verlag, pp. 381–98.
Ong, W.J. (1982), *Orality and Literacy. The Technologising of the Word*, London, Methuen.
Park, R.E. and Burgess, E.W. (1924), *Introduction to the Science of Sociology*, 2nd edition, Chicago, University of Chicago Press.
Parkes, D.N. and Thrift, N.J. (1980), *Times, Spaces, Places. A Chronogeographic Perspective*, Chichester, Wiley.
Philip, A. (1921), *The Calendar*, Cambridge, Cambridge University Press.
Philo, C. (1986), 'Geography, history and the mad-business', unpublished manuscript, Department of Geography, University of Cambridge.
Platt, C. (1969), *The Monastic Grange in Medieval England. A Reassessment*, London, Macmillan.
Price, P. (1983), *Bells and Man*, Oxford, Oxford University Press.
Reid, D.A. (1976), 'The decline of Saint Monday, 1766–1876', *Past and Present* (May), no. 71, pp. 76-101.
Rodwell, W. (1981), *The Archaeology of the English Church. The Study of Historic Churches and Churchyards*, London, Batsford.
Rowley, T. (1983), *The Norman Heritage 1066–1200*, London, Paladin.
Salzman, L.F. (1967), *Building in England Down to 1450*, (second impression), London, Oxford University Press.
Scott, J.G.M. (no date), *The Bells of Exeter Cathedral*, Exeter.
Smith, T.C. (1986), 'Peasant time and factory time in Japan', *Past and Present*, no. 111 pp. 165–97.
Steane, J.M. (1984), *The Archaeology of Medieval England and Wales*, London, Guild Publishing.
Thomas, K. (1971), *Religion and the Decline of Magic*, London, Weidenfeld & Nicolson.
Thompson, E.P. (1967), 'Time, work-discipline and industrial capitalism', *Past and Present*, no. 38, pp. 56–97.
Thompson, J.W. (1963), *The Literacy of the Laity in the Middle Ages*, New York, Burt Franklin.
Thrift, N.F. (1977), *An Introduction to Time–Geography*, CATMOG 13, Norwich, Geo Abstracts.
Thrift, N.J. (1981), 'Owners' time and own time: the making of a capitalist time consciousness, 1300–1800', in A.R. Pred (ed.), *Space and Time in Geography*, Lund, Gleerup, pp. 56–84.
Thrift, N.J. (1985), 'Flies and germs: a geography of knowledge' in D. Gregory and J. Urry (eds.), *Social Relations and Spatial Structures*, London, Macmillan, pp. 366–403.

Thrift, N.J. (1986), 'Little games and big stories: accounting for the personality of political practice in the 1945 General Election' in K. Hoggart, and E. Kofman (eds), *Politics, Geography and Social Stratification*, Beckenham, Kent, Croom Helm, pp. 86–143.

Treharne, R.F. (1971), *Essays on Thirteenth Century England*, London, Historical Association.

Tribe, K. (1981), *Genealogies of Capitalism*, London, Macmillan.

Whipp, R. (1987), ' "A time to every purpose": an essay on time and work', in P. Joyce (ed.), *The Historical Meanings of Work*, Cambridge, Cambridge University Press.

White, L. (1962), *Medieval Technology and Social Change*, Oxford, Clarendon Press.

Yates, F.A. (1966), *The Art of Memory*, Chicago, University of Chicago Press.

Zerubavel, E. (1981), *Hidden Rhythms*, Chicago, University of Chicago Press.

CHAPTER 5

TIME AND WORK ORGANISATION: A THEORETICAL AND EMPIRICAL ANALYSIS

Ken Starkey

INTRODUCTION: TIME AND WORK

Time is a social experience and its primary determinant is the experience of work. 'Work provides the guidelines by which adults ordinarily structure their use of time' (Calkins, 1970, p. 487). There is a developing research tradition, called by Fraser (1968) 'chronosophy', which informs this view. In the field of chronosophy a new research agenda is being created focussing on 'the sociotemporal order' (Zerubavel, 1981, p. xii), the temporal order that is created by the way social groups experience and manage, or attempt to manage, their time. A fundamental contention of this research is that time has for too long been neglected in the analysis of the social order (Lüscher, 1974), that social life is structured and regulated at its most basic level by its temporal parameters, yet that the social sciences have, with notable exceptions, taken time for granted. The aim of current research in the field of social time is to question this 'taken-for-grantedness' in demonstrating both how 'concepts of time . . . facilitate the reproduction of social relations . . . yet are highly problematic' (Clark, 1982, p. 9).

Fraser sees as central to chronosophy a comparative analysis of historical and current ideas about time, the analysis of the role of time in the evolution of societies and the examination of ways of reasoning about time (Fraser, 1968, p. 592). Durkheim emphasised the relativity of social time. In *The Elementary Forms of the Religious Life* he argued that the origins of time were social, drawing his conclusions from the evidence of studies of primitive societies, and concluding that 'the foundation of the category of time is the rhythm of social life' (Durkheim, 1976, p. 22). He was thus the first to focus on the collective social aspect of temporality. Time is not, as Kant held, a universal category inherent in the mind. It is essentially a 'collective representation', arising from the experience of the collectivity. Society provides the framework according to which time is arranged. Primitive societies arrange this framework to correspond to the 'periodical recurrence of rites, feasts, and public ceremonies' (Durkheim 1976, p. 32). Hubert and Mauss (1909) make similar points in their study of time in religion and magic in which the representation of time is clearly revealed in its qualitative rather than its quantitative aspects. This aspect of social time was most thoroughly examined by Sorokin in his contrasting of sociocultural (qualitative) and sidereal (quantitative) time (Sorokin, 1943). The important lesson for the social scientists who succeeded and whose work was informed by these founding fathers of the discipline of chronosophy is that, as Zerubavel expresses it:

> Despite the growing prevalence of the quantitative conception of temporality . . . people clearly do not relate to the time only as a physico-mathematical entity. They also view it from a qualitative perspective, as an entity which is imbued with meaning. (Zerubavel, 1981, p. 101)

Time, then, is more complex than the dominant homogeneous quantitative concept indicates. It is social in origin. Anthropology demonstrates that different cultures develop different notions of temporality. Shared conceptions of time make social life possible. 'Time is the creation of the group for the life of the group. It is a source of regularity and dependableness' (Bosserman, 1975a, p. 23). As cultures evolve so do their conceptions of time and, also, different social groups in the same culture come to experience and orientate themselves towards time in differing ways. This facet of time has been explored by Gurvitch and Mannheim. Mannheim describes different social groups' aspirations in terms of their conceptions of time which, he argues, demonstrate most clearly 'the innermost structure of the mentality' of a group (Mannheim, 1936, p. 188). Gurvitch creates a typology of social times, a spectrum of their features, and points out, for the first time, that different groups exist in divergent times which are often in competition with each other (Gurvitch, 1963, p. 174). Social classes have radically differing conceptions of time, different, for example, in terms of duration (temporary, enduring, or permanent) or rhythm (slow, medium or rapid). A central problem of the social theorist is to demonstrate which segment of society 'dominates and controls the time scale' (Gurvitch, 1963, p. xxvi). Dominant in Western industrial democracies is a commodified conception of time, epitomised in the saying 'time is money'. A consequence of such a conception is the quantification of time and efforts to organise social time in terms of economic production and profit. It is always too short for management and too long for labour. Management tries to master time in order to prolong it while labour tries to master time in order to shorten it (Gurvitch, 1963, p. 44).

The centrality of a quantitative view of time has grown with the progress of industrial organisation and has been facilitated by the application of clock time to social organisation. Mumford (1973) emphasises this point when he stresses that the first characteristic of modern machine civilisation is its temporal regularity and that this depends for its precise regulation on the clock. The clock is the 'crucial' machine of an industrial society (Moore, 1963, p. 163) and the management of time is a major problem of industrial organisation. Weber (1976) isolates the rationalisation of time as a fundamental, if not the fundamental characteristic of the Spirit of Capitalism. Time is the primary scarce resource, the use of which has to be maximised. It is used as a commodity, to be traded like any other. The industrial labour process had to replace 'the natural time order of earlier agrarian societies' (Littler and Salaman, 1984, pp. 53–5) with what E.P. Thompson called a 'task-orientation' to time (Thompson, 1967). Before industrialisation the task was more important than time. Industrialisation reversed the relationship. A new time orientation and time discipline were enforced. The discipline has to be constantly reproduced (Littler and Salaman, 1984, p. 54; Willis, 1977). Time is a key element of industrial control.

The commodification of time is highlighted by Marx as a guiding principle of capitalism.

> The underlying constitutive component of both goods and labour that permits their common existence as interchangeable commodities, according to Marx, is *time*. Every commodity, including labour-power itself, is 'the objectivication of a given amount of labour time'. The 'socially necessary labour time' governs the value of commodities, and is the standard measure of exchange-value. Units of time are what makes the value of commodities divisible and quantifiable. The quantification of time is thus that very foundation of the universalising character assumed by the exchange of commodities. In economic calculation, accounting, and so on, commodities are transformed into quantitative measures of value on paper. On paper, as Marx says, this process proceeds by mere abstraction; but 'in the real exchange process, a real *mediation* is required, a means to accomplish this abstraction'. *This means is supplied by the fact that commodities in exchange exist only as exchange-values, which in turn depends upon the temporal equation of units of labour.* (Giddens, 1981, pp. 118–19)

One of the key themes of *Capital* is that 'the economic order of capitalism depends upon the exact control of time' (Giddens, 1979, p. 210). The regulation, and exploitation, of labour time is central to the economic development of the capitalist system. The logic of capitalist development entails the maximisation of productivity in time either by the lengthening of the working day or, when this becomes impossible, by the intensification of the maximally available time. Time becomes everything, 'man is nothing; he is, at most, the carcase of time'. This, Marx proclaims, is the central fact of modern industry (Marx and Engels, 1976, p. 127). As much of the lifetime as possible of the labouring class is converted by the economic pressure of the capitalist class into labour time. No other time is important. Management aims to reduce to a minimum the 'porosity' of working time, filling the 'pores' of empty time with productive labour (Palloix, 1976; Clark et al., 1984).

Time is linked to the Marxist concept of alienation and alienated labour. Work becomes

> the more alien the more specialized the division of labour. Men do not live their own lives but perform pre-established functions. While they work, they do not fulfill their own needs and faculties but work in *alienation*. Work has now become *general*, and so have the restrictions placed upon the libido: labour time, which is the largest part of the individual's life time, is painful time, for alienated labour is absence of gratification, negation of the pleasure principle. (Marcuse, 1969, p. 51)

Marcuse links this process to the 'temporal reduction' of the libido. Time is 'the deadly enemy of Eros' (Marcuse, 1969, pp. 52–3, 186). Radical Marxists such as the situationists see time as an octopus crushing its prey. Time under capitalism is the 'gigantic pincer with which power torments humanity' (Vaneigem, 1967, p. 156). Sève elaborates the Marxist critique of the dominant time perspective to incorporate a Marxist theory of personality, bringing together all these themes in the analysis of the 'sacrifice of concrete personal life to abstract social life' regulated by abstract time (Sève, 1978, p. 343). Commodified time necessitates the allocation of time to separate often contradictory activities. The Marxist theory of personality examines these contradictions in terms of 'the ensemble of processes of separation and partitioning between [the] different sectors' of the personality (Sève, 1978, p. 362).

A key theme in Marx is the struggle for the control of time exemplified in the struggle for time free from work over which the worker exercises his own control. Alienation as self-estrangement arises because the worker does not have the time for self-development. Ricardo's postulate that wealth is nothing more than disposable time is agreed with and disposable time contrasted with the coercion of labour time. 'The individual is seen to be at his highest level of development when the dichotomy between necessary labour and disposable time, that is free time, no longer exists' (Walliman, 1981, p. 109). It is the worker's time which is at the core of the creation of surplus value and, therefore, exploitation, insofar as the worker is not fully recompensed for the time devoted to work. This is the central tenet of the labour theory of value. A prime goal is to increase the freedom from the domination of working time. With the abolition of the capitalist mode of production and the establishment of communism there will be a reduction of labour to a minimum and a maximisation of free time.

TIME AND WORK ORGANISATION

Shared concepts of and ways of mutual interaction in time are essential to social order and to the survival of any organisation. Regularity of shared performance depends upon them. Organisations depend upon the synchronisation of activity, its correct sequencing and an agreed rate of work (Moore, 1963, p. 163). Organisational survival depends upon the ability to organise the routine. Work discipline focuses upon correct attitudes to time and individual socialisation into an organisation is comprised in large part of adaptation to the demands of organisational time. For the newcomer time is problematic: 'little can be accomplished . . . of organizational interest unless [the newcomer] can locate himself along the . . . axes of organizational time and space' (van Maanen and Katz, 1979, p. 33).

Just as time is limited as a personal resource (Becker, 1977), organisations usually operate under conditions of time scarcity. Time use, therefore, has to be maximised. Work relations are maintained and reproduced by time-discipline, initially imposed by force but now, more usually, internalised during the processes of socialisation (Foucault, 1979; Willis, 1977). Simmel (1971) stresses punctuality, calculability and exactness as key social schemata, inherent in existing social relations and essential to the continuing existence of these relations. Socialising factors in early childhood reinforce these schemata (Piaget, 1969) and education continues the process, further inculcating the temporal habits upon which acquiescence to the demands of work time will depend

(Willis, 1977). Activity is regulated by the timetable which divides the day into a sequence of set tasks. Key managerial work organisation tasks are control and coordination.

Work in industrial civilisation has increasingly depended upon compliance with the clock, but compliance has not always been peacefully negotiated. Time has been an important arena of contestation (Thompson, 1967). The control of time has long been and continues to be a key industrial relations issue as struggles over the length of the working week and for free time demonstrate. It is a central Marxist argument that as hours of work decrease management compensates by intensifying the pace of work, employing such means as stricter supervision, incentive payment schemes and technological innovation. The progress of industrialisation saw the imposition of ever more demanding work timetables and consequent conflicts over time in which such innovations as time-and-motion study were resisted by go-slows, other more subtle ways of controlling output, and by absenteeism and strikes. The two main contributions to the evolution of managerial control systems – scientific management, based on the work of F.W. Taylor, and Fordism, based on the management practices of Henry Ford – have at their roots a concern with working time. Scientific management aimed at maximising the use of working time by emphasising the 'one best way' of doing work and by eliminating such phenomena as 'systematic soldiering' by which workers organised collectively to restrict output and control earnings (Taylor, 1911). Fordism intensified work by introducing new forms of technical control, embodying the control of the pace of working in the speed of the assembly line (Edwards, 1979). The assembly line dictated the pace and therefore the intensity of work (Linhart, 1981). The clock becomes 'the key machine of the modern industrial age . . . a piece of power machinery' (Mumford, 1973, pp. 271–72; Giddens, 1979; Clark 1982).

The conflict of classes around economic interest at work promotes a continuous search by management for ways to control and cheapen the labour process (Child, 1985). This search, exemplified in different forms of work organisation, impacts on time in two of its aspects: duration and intensity. With reductions in the length of the working week management has become primarily concerned with the second of these. The managerial imperative is the intensification of time use, the maximisation of productive labour through the most productive possible use of time. Labour intensification is a prime component of the rationalisation of work organisation which aims at 'ever more and more corporate influence over the direction, quantity, quality and intensity of workers' effort' (Littler and Salaman, 1984, pp. 35–6). Labour intensification is defined as 'an increase in the proportion of the working day consumed in actually working', i.e. in terms of 'effort level', 'an increase in the rate of working (rate of effort expenditure)' (Kelly, 1982, p. 196).

> More intensive working may take the form either of *faster* working, i.e. an increased rate of effort expenditure in the same time, or of more prolonged working within the working day. Where productivity is raised in this way we can speak of *intensification* of labour. (Kelly, 1982, p. 59)

TIME AND THE RAILROADER

This section examines the themes of time and intensification as exemplified in the case of railway drivers and 'flexible rostering'. In 1982 a bitter dispute erupted on British Rail between the British Railways Board (BRB) and the Associated Society of Locomotive Engineers and Firemen (ASLEF). The dispute concerned the introduction of 'flexible rostering', a new pattern of working organised around shifts varying in length between seven and nine hours, to replace a shift system organised around a standard eight-hour day. ASLEF rejected the new shift pattern, arguing that it meant a disruption of established working routines; that it required a wide range of shift starting times, often in the same week; that drivers would have a wide range of weekly working hours with less regular free weekends; that there would be great difficulty in swapping the new irregular shifts when personal need required time off work; and that, as a result, drivers' social and family lives would be exposed to an even greater degree of stress than they already experienced because flexible rostering would mean an increasing dominance in their lives of working time and would reduce their freedom to use their non-work time as they saw fit. ASLEF resisted the imposition of flexible rostering by British Rail, going on strike in the summer of 1982. The union was forced to call off strike action and return to work when the Trade Union Congress refused to back their case. Railway drivers' work is now organised according to principles of flexible rostering.

Time has always been a central focus of railway culture. As long ago as 1939, Cottrell, in his seminal paper 'Time and the railroader', argued that the importance of time to the efficient running of the railway system deeply marked the whole of the railroader's life, working and non-working. Consequently the railway worker has been subjected to as unrelenting a time discipline as any other group of workers. The rigidity of the timetable, the system's subconscious mind, and the unrelenting time discipline it demanded, created an overarching scheme that controlled all of the railroader's life. The railroader became

> a slave to the clock; intense time-consciousness marks the railroader in all his social relationships . . . The obverse of this occupational dependence on the clock is the almost complete denial of the opportunity to time-plan other relationships. (Cottrell, 1939, pp. 195–6)

Kemnitzer, reassessing Cottrell's work, draws broadly similar conclusions, while stressing the growing importance of 'accurate mechanical time' as more than a mere 'operating tool'. It is now highlighted as the key element in the competition for profit (Kemnitzer, 1977, p. 25). Clock time has become even more important because of this. Kemnitzer also argues that there has been one fundamental change in railroader attitudes since the Cottrell study was published. He no longer identifies with the system's temporal values, an effect of broad cultural changes that tend to emphasise the importance of identification with non-work roles (Goldthorpe et al., 1968) and the increased centrality of leisure time (Dumazedier, 1967). Gamst's ethnographic study of footplate work (1980) paints a similar picture to Cottrell's and Kemnitzer's.

The work of Cottrell, Kemnitzer and Gamst refers to the American railroad system. However McKenna, a former ASLEF trade union official, makes broadly similar points concerning the evolution and effects of time discipline on railway work in the UK, ending his book with a chapter entitled 'The tyranny of time', discussing the 'clock-and-watch-bound bondage' of the railway worker's 'roped existence' (McKenna, 1980, p. 250). It seems probable that the American 'tightening-up' on time would be more rigorous than the British as the British form of capitalism has not led to as strong a valorisation of time as has occurred in the USA (Hall, 1959). American management demonstrates a more strongly developed sense of time as a key commodity and is, therefore, more sensitive to different ways of organising time (Clark, 1987). But what McKenna highlights is a similar reaction on British management's part to problems of time management. Struggles over the length of the working week led to a comparable intensification of time use through such innovations as the big engine policy – 'bigger, longer, heavier, faster'. If time itself could not be made available in bigger, longer quantities, it could be employed more productively. Time, thus, comes to be experienced as faster and weighing more heavily on the worker operating according to new temporal demands, support for the Marxist argument concerning the intensification of the pace of work.

This raises the issue of how working time is to be organised most productively if it cannot be extended through time *ad infinitum*. Recent developments on British Rail, namely the introduction of flexible rostering, indicate that the locus of work organisation debates concerning time has shifted from the length of the working week – how far work should extend through time – to the intensity of a reduced number of hours – how working time is to be organised to maximise the intensity of use of a limited time period. This, I would argue, is the key industrial relations time issue of the future. The British Rail Board – ASLEF dispute concerning flexible rostering and the intensification of the use of time can be seen as a paradigm for future struggles over the reorganisation of time and the effects it is seen by the workforce as having on their temporal situation. The focal point of disputes about time is seen to have shifted, in this case, from the dimension of duration to the dimension of intensity. Cottrell concentrated on the amount of time which railroad work demanded but much of this involved waiting for the next piece of activity and doing nothing in between. With gains in non-work time accomplished, the primary arena of struggle shifts to the quality of the time spent on the job and how this is to be best organised to suit management and labour needs. In the flexible rostering dispute, ASLEF construed proposed alterations to the work roster system as a significant deterioration in their working situation, brought about by an intensification of work. The dispute marks a shift away from the more traditional quantitative focus on duration of the working week to the issue of time as a qualitative concern for both management and labour.

Time discipline was the focal issue in the BRB–ASLEF dispute concerning flexible rostering, one of the bitterest industrial disputes of recent years, described by the Railway Staff National Tribunal (the RSNT), whose task it was to adjudicate in the dispute, as 'the most difficult reference we have had to deal with' as the issues were so complex and far-reaching for the future of the railways (RSNT, 1982, para. 77). There were three strands to the dispute – the implementation of a pay rise, the reduction of the working week to thirty-nine hours, and the introduction of flexible rostering for footplatemen. It was agreed by both parties in the dispute that the third issue was of paramount importance. When ASLEF refused to cooperate in the introduction of flexible rostering, which would have involved relinquishing a long-standing agreement that there should be a standard eight-hour working day and the acceptance of some flexibility around the eight hours, with rosters of up to nine hours being allowed, BRB refused to implement a previous pay and length of working week agreement. ASLEF responded with strike action and the Advisory, Conciliation and Arbitration Service (ACAS) intervened. Here the points at issue were set out, ASLEF's argument being put in their absence as they refused representation.

ASLEF was presented to ACAS as intransigent in its opposition to any change in the guaranteed eight-hour day, sacrosanct for footplatemen since its agreement in 1919. BRB was equally intransigent in its opposition to the eight-hour day's rigidity. Flexibility of hours was claimed to be essential to more effective use of working time, to 'increasing efficiency, productivity and offsetting the cost effect of a shorter working day' (ACAS, 1982, para. 20). BRB also argued that flexible rostering would work in favour of reducing unsociable hours by introducing more rest days, longer periods clear of duty and by lessening the need to sign on between midnight and five a.m. This would be to the advantage of employees as would the overall reduction in working time. ASLEF argued, or, rather, ACAS argued for them in their absence, that the gist of their case against flexible rostering was that it would exacerbate the already substantial stresses and strains that footplate work was especially prone to. The outcome of the ACAS meeting was a call for a more detailed discussion of the issue at the RSNT whose role it is to decide on issues pertaining to standard salaries, wages, hours of duty and other conditions of railway service. This meeting, which ASLEF did attend and submit detailed evidence to, gave rise to RSNT Decision No. 77.

The core of the BRB's case to the Tribunal was the obsolescence of an agreement negotiated sixty-seven years previously, and at issue was 'whether BR would be tied indefinitely to agreements which limit its efficiency and which could be changed without overall disbenefit to the workforce' (RSNT, 1982, para. 33). It argued that only a minority of rosters would require longer hours than those presently worked and that there would be no increase in overtime or in the amount of unsocial hours. ASLEF claimed all three of these would occur. Where BRB agreed ASLEF might have a case – the only point of agreement in the evidence the two sides submitted – was in the arrangement of exchanges of turns of duty. Non-standard time-slots might create problems in this respect. BRB also argued that footplatemen's control over rostering arrangements would also be improved by an envisaged increased participation at the local departmental committee level in their construction.

BRB claimed that more flexibility of time management was necessary as the timetable itself could not be neatly divided into eight hour pieces 'or any other size of pieces: no amount of planning could make it do so within the present rigid arrangements' (BRB, 1982, para. 40) due to the rigid framework resulting from ancient agreements. The average working time on an eight-hour shift was only three hours twenty minutes. Even a ten minute increase in the working content would represent a five per cent improvement (BRB, 1982, para. 40). With a thirty-nine instead of a forty-hour week the cost of each hour of work would increase by 2.56 per cent. Minimum cost implementation of a shorter working week depended on performing the required work in less hours.

These were BRB's economic arguments for flexible rostering. ASLEF's opposition focussed on two lines of argument: the implications of flexible rostering for the future of railway jobs and the psycho-physiological effects of the proposed changes in time practices. BRB had not disguised its quest for the rationalisation of jobs but stated that its effects would be long- rather than short-term and would not result in compulsory redundancies. This ASLEF objected to in principle but the main thrust of their case centred on the effects time reorganisation would have on their present members' lives. BRB, they argued, in its preoccupation with reducing the wages bill, was ignoring the social implications of its proposed changes. Footplate work was unique in its demands for attention and concentration, the hours of work were already highly irregular and unsocial and the existing eight-hour day constituted the only fixed point in a working pattern of extreme flexibility. Exceeding eight hours of single-manning a train could take drivers beyond the limits of psycho-physiological tolerance and would certainly result in an increase in the stress level of the work. The driver's social life would be made 'virtually impossible' (RSNT, 1982, para. 69) as hours of work became even more irregular.

It was BRB's fault if it could not roster efficiently using the existing eight-hour time-slots (ASLEF, 1982, para. 35). Even with the eight-hour day there were already instances where drivers were away from home for twelve hours or more. Greater flexibility could only exacerbate this. Worst of all, ASLEF suspected that the presently proposed form of flexible rostering was only the beginning:

> . . . it is clear that BR's ultimate aim is to carry this principle much further with an eventual flexibility to 4–10 or even 4–12 hours, the principle being that drivers whose lives are already greatly disturbed to fit the needs of the railway service (sometimes necessarily so) put their own time almost entirely at the disposal of the BRB in every respect. (RSNT, 1982, para. 116)

The hidden agenda in ASLEF's case is the spectre of a return to a time commitment similar to that demanded by the feudal baronetcy of the Victorian era, a fear that impending changes presaged the railroader becoming, once again, the slave of railway time. Flexible rostering would increase the centrality of working time in the railway driver's life and its acceptance could only involve a retrograde step for the workforce.

The argument over the control of time was central to the dispute. The argument concerned quantity and quality of time. There was the dispute about the parameters of clock time, the total amount of labour time to be paid for and what percentage of that time should be available for work and in what time form, the argument about how to make more than the previous average of three hours twenty minutes available. A reduced number of hours was to be used more intensely. It was this intensification of its time usage that the union resisted and the psycho-social effects it felt the changes would bring in their wake. The union opposed the social effects of an economic approach to time allocation. BRB, however, suspected that ASLEF resistance to changing work practices was, at root, also economically based, that prevarication actually constituted a ruse aimed not at the amelioration of working conditions but at gaining a substantial increase in pay by claiming a guaranteed minimum payment for eight hours, as the price for accepting it:

> [ASLEF] has more than once . . . said that they would be willing to consider programs up to nine hours providing *all* turns carried a minimum guaranteed payment of eight hours . . . Judging by their statements they are less concerned with the length of turn than with the way they

will be paid for working it. (BRB, 1982, para. 69)

BRB could argue, therefore, that what was involved was, fundamentally, a single view of time, that both sides shared a commodified view of time as money and were manoeuvring to increase their share. Time was the pawn in a game that was really about money. ASLEF's case rested on the view that there are two ways of looking at time and that the social, qualitative way should have precedence over the quantitative and economic. The management concept incorporates notions of efficiency, productivity, and the rationalisation of time use by intensification based on more flexible allocation of time to work tasks. The union did not reject these aims. It argued, rather, that the social consequences of their implementation had to be taken into account. For ASLEF a certain form of time – the eight-hour day – was sacrosanct. Its inception had marked a major gain for the union and its retention was of symbolic significance. The union could not visualise a time-frame that was not based on this and feared the worst. As Ray Buckton, ASLEF General Secretary, expressed it: 'it is my Society's considered opinion that the railway industry has not, as yet, seen flexible rostering in operation' (personal communication). The union accepted the importance of work time but also stated that there are other times, most notably family and leisure time, that have to be taken into account. Work should not dominate, even to the extent it already did before flexible rostering. There should be 'as much emphasis on leisure time as the need to earn money to live', it had stated in a previous argument for a shorter working week (RSNT, 1979, para. 38). The previous time situation had been unsatisfactory in quantitative and qualitative terms. Time was already too long, too fast, too full, and flexible rostering would make it worse.

CONCLUSION

Work organisation has to be understood as a complex of social, technical and economic factors in a context that includes external pressures on organisations as well as internal structures and processes (Clark and Starkey, 1987).

Social, economic and political factors all impact on managerial strategies. In private sector firms the pressure of the market has been emphasised as a key constraint (Kelly, 1982; Child, 1985). In the public sector the situation is more complex although the 'Thatcherite' response to the 'fiscal crisis of the state' (O'Connor, 1975) has seen the diffusion of private sector managerial goals to the public sector. Ferner (1985) has analysed those determinants of BR's industrial relations strategy which are peculiar to a nationalised industry. Political constraints determine 'the sorts of objectives and goals [state firms] are expected to pursue and the rules according to which they pursue them' (Ferner, 1985, p. 47). Political constraints are particularly important to a company like British Rail which is so heavily dependent on state subsidy. The fiscal crisis of the state formed the 'backcloth' to the productivity initiative of the BRB. The Conservative government's attitude to the public sector and the railways in particular made the political pressures all the more acute (Serpell, 1983).

The focus of political pressure and the nature of state intervention vary from industry to industry according to a complex set of factors.

> These include the nature of the industry, the type of political interests (both public and private) that mobilise around it (including suppliers, competitors, customers, state administrators and sectors of the labour movement), and internal factors such as the forms of accommodation between management and workforce or unions. (Ferner, 1985, p. 67)

But if the focus and the nature of these interventions varies, the goal of management remains the same. Batstone et al. conclude their study of public sector postal and telecommunications services: 'in both public and private sector, management will at times be forced – whether by political contingency or by the dictates of a crisis – to exert greater control over work organisation' (Batstone et al., 1984, p. 293). Greater control will aim, ultimately, at the intensification of work and the maximisation of the use of labour time. It is for this reason that the study of time is so important to our understanding of work organisation. According to a Marxist perspective capitalism and managers as representatives of capital proceed according to the principle of an ever more intensive use of time, aiming to extract from time 'ever more available moments and from each moment ever more useful forces' (Foucault, 1979, pp. 151–2). The arguments for flexible rostering are based on this principle. From the Marxist perspective ASLEF's refusal to accept flexible rostering represents a radical refusal of the principle of intensification of time use. It constitutes a fundamental rejection of a capitalist order based on the commodification and control of time and the defence of the union's right to the degree of temporal autonomy ratified by long-standing agreements. The other way of conceptualising ASLEF's position is in economistic terms, that ASLEF's refusal constitutes not a rejection of the capitalist rules of the game regarding time but a ruse to maximise the financial return on time worked. Thus BRB argued that resistance to flexible rostering was a ploy in the bargaining process aimed at extracting more money in later negotiations as the price of accepting the new system.

These contrasting interpretations of ASLEF's actions raise important questions about the rules according to which unions negotiate time. The management case rests squarely on a commodified approach to time. The union case is more problematic. It can be construed as a rejection of the dominant managerial time ideology – the view that time, essentially, is money – in its appeal to non-economic criteria concerning the psycho-social effects of the intensification of time-use. It can also be construed, in similar vein, as a rejection of the dominant ideology concerning the power to manage time. In this respect the flexible rostering dispute is further evidence of the lack of normative consensus that bedevils British industrial relations (Hill, 1981). The lack of consensus concerns the degree of time commitment workers and management consider equitable. The union argues that the extent of permissible time management has been long agreed and established by custom and practice. Management argues that traditional ways of organising time are no longer efficient enough, given the precarious state of BR finances in the current economic climate. Divergence on this point represents a radical split in temporal values. On the other hand, it can also be argued that ASLEF is not rejecting the commodification principle espoused by management *per se* if the union's case is seen as support for the view that the present return in time worked for the wages paid is a just one and that more time worked will need to be bought with a wage increase. The union's case seems to be balanced between these two positions, the one anti-capitalist in its rejection of the managerial view of time as nothing more than money, the other economistic. The two positions counterpoint each other in the dispute with the union perhaps uncertain about which to pursue. If the former is a path followed in future disputes concerning negotiations about time then other issues besides the specification of an equitable wage will come to the fore, issues concerning the most human as well as the most efficient way to organise work time.

NOTE: TIME AND SKILL

An undercurrent in the ASLEF–BRB dispute, not explored in this paper, is the relationship between the management of work time and the exercise of work skills and here the spectre of deskilling enters the debate as a possible consequence of intensification. The 'normative basis to levels of work effort must be socially constituted' (Littler, 1982, p. 61). Standard effort levels and, therefore, standard ways of structuring time are 'built into' occupations and the skills that constitute these occupations. If these customary standards are disrupted it impacts on the skill level of the work. In this respect, it is important to remember that ASLEF is a craft union whose identity rests on a particular concept of skill. The intensification of work through flexible rostering strikes at the very basis of skill, the possibility of the proper exercise of which has its temporal roots in the long-established custom of the eight-hour day. A concentration of time use will dilute the possibility of exercising skills fully by demanding a commitment of energy beyond the level the union considers possible. The exercise of skill cannot be maintained for the more prolonged periods that longer days will necessitate. The mechanical conceptualisation of time upon which commodification of time rests does not take this view into account. It does not relate skill and effort to time but assumes that if a task can be performed its prolongation past those periods tested by custom and practice is unproblematic. In proposing an intensification of time use BRB assumes that skill levels can be maintained regardless of time but the exercise of a skill for more than an optimum period dilutes the energy necessary to sustain the exercise of that skill. ASLEF rejects the time–effort–energy relationship that BRB does not recognise as a potential problem. A craft union rejects a principle of work organisation that might be applicable to work with lower skill levels without deleterious effects on the exercise of skill. Here it is interesting to remember that the National Union of Railwaymen (the NUR) accepted flexible rostering as unproblematic. An analysis of this acceptance would need to deal with the differences between industrial and craft trade union attitudes to time and the relevance of the concept of skill as the preserve of an occupational group that Bagwell (1982) raises in his study of the complex relationship between ASLEF and the NUR.

REFERENCES

ACAS (Advisory, Conciliation and Arbitration Service) (1982), 'Committee of Inquiry Report and Recommendation on a Dispute between the British Railways Board and the Associated Society of Locomotive Engineers and Firemen', London, ACAS.

ASLEF (Associated Society of Locomotive Engineers and Firemen) (1982), 'Submission by R.W. Buckton, the General Secretary to the Railway Staff National Tribunal', London, ASLEF.

Bagwell, P.S. (1982), *The Railwaymen*, vol. 2, *The Beeching Era and After*, London, Allen & Unwin.

Batstone, E. et. al. (1984), *Consent and Efficiency: Management Strategy and Labour Relations in the State Enterprise*, Oxford, Blackwell.

Becker, G.S. (1977), *The Economic Approach to Human Behaviour*, Chicago and London, University of Chicago Press.

Bergson, H. (1910), *Time and Free Will. An essay on the immediate date of consciousness*, London, Allen & Unwin.

Bosserman, P. (1975a), 'Some theoretical dimensions to the study

of social time', *Society and Leisure*, vol. 7, pp. 21–44.
Bosserman, P. (1975b), 'The evolution of trends in work and non-work time in United States society, 1920–1970', *Society and Leisure*, vol. 7, pp. 89–132.
Braverman, H. (1974), *Labor and Monopoly Capital: the Degradation of Work in the Twentieth Century*, New York and London, Monthly Review Press.
BRB (British Rail Board) (1982), 'Proposals for the implementation of sub-paragraph 2(c) of the 1981 productivity understanding to footplate staff' (Submission to Railway Staff National Tribunal), London, British Rail Board.
Calkins, K. (1970), 'Time: perspectives, marking and styles of usage', *Social Problems*, vol. 20, pp. 487–501.
Child, J. (1985), 'Managerial strategies, new technology and the labour process', in D. Knights et al. (eds), *Job Redesign. Critical Perspective on the Labour Process*, Aldershot, Gower.
Clark, P. (1982), 'A review of the theories of time and structure for organizational sociology', to appear in S. Bachrach (ed.), *Perspectives in Organization Sociology: Theory and Research*, vol. 2, United States of America, J.A.I. Press.
Clark, P., Hantrais, L., Hassard, L., Linhart, D. and Starkey, K. (1984), 'The porous day', paper presented to the Aston/UMIST Labour Process Conference, Aston University.
Clark, P. (1987), *Anglo-American Patterns of Commercial and Industrial Innovation*, London, De Gruyter.
Clark, P. and Starkey, K. (1987), *Organizational Transitions and Design-Innovation*, London, Frances Pinter.
Clutterbuck, D. (ed.) (1985), *New Patterns of Work*, Aldershot, Gower.
Cottrell, W.F. (1939), 'Of time and the railroader', *American Journal of Sociology*, vol. 4, pp. 190–98.
Dumazedier, J. (1967), *Towards a Society of Leisure*, New York, Free Press.
Durkheim, E. (1976), *The Elementary Forms of the Religious Life*, 2nd edition, London, George Allen & Unwin.
Edwards, R. (1979), *Contested Terrain: the Transformation of the Workplace in the Twentieth Century*, London, Heinemann.
Ferner, A. (1985), 'Political constraints and management strategies: the case of working practices on British Rail', *British Journal of Industrial Relations*, vol. 23, pp. 47–70.
Foucault, M. (1979), *Discipline and Punish*, Harmondsworth, Penguin.
Fraisse, P. (1964), *The Psychology of Time*, London, Eyre &

Spottiswoode.
Fraser, J.T. (ed.) (1968), *The Voices of Time*, Harmondsworth, Penguin.
Gamst, F.C. (1980), *The Hogshead: an industrial ethology of the locomotive engineer*, New York and London, Holt, Rinehart & Winston.
Giddens, A. (1979), *Central Problems in Social Theory*, London, Macmillan.
Giddens, A. (1981), *A Contemporary Critique of Historical Materialism*, vol. 1, *Power, Property and the State*, London, Macmillan.
Goldthorpe, J.H. et al. (1968), *The Affluent Worker: Industrial Attitudes and Behaviour*, Cambridge, Cambridge University Press.
Gurvitch, G. (1963), *The Spectrum of Social Time*, Dordrecht, D. Reidel.
Hall, E.T. (1959), *The Silent Language*, Garden City, Doubleday.
Hill, S. (1981), *Competition and Control at Work*, London, Heinemann.
Hubert, H. and Mauss, M. (1909), 'Etude sommaire de la representation du temps dans la religion et la magie', in *Mélanges d'histoire des religions*, Paris, Alcan, pp. 199–208.
Kelly, J. (1982), *Scientific Management. Job Redesign and Work Performance*, New York, Academic Press.
Kemnitzer, L.S. (1977), 'Another view of time and the railroader', *Anthropological Quarterly*, vol. 50, pp. 25–9.
Linhart, R. (1981), *The Assembly Line*, London, Calder.
Littler, C.R. (1982), *The Development of the Labour Process in Capitalist Societies*, London, Heinemann.
Littler, C.R. and Salaman, G. (1984), *Class at Work. The design, allocation and control of jobs*, London, Batsford.
Lüscher, K.K. (1974), 'Time: a much neglected dimension of social theory and research', *Sociological Analysis and Theory*, vol. 4, no. 3, pp. 101–17.
Maanen, J. van and Katz, R. (1979), 'Police perceptions of their environment', *Sociology of Work and Occupations*, vol. 6, pp. 31–58.
McKenna, F. (1980), *The Railway Workers 1840–1970*, London, Faber.
Mannheim, K. (1936), *Ideology and Utopia: an introduction to the sociology of knowledge*, London, Routledge & Kegan Paul.

Marcuse, J. (1969), *Eros and Civilization*, London, Sphere Books.
Marx, K. and Engels, F. (1976), *Collected Works*, vol. 6, London, Lawrence & Wishart.
Moore, W.E. (1963), 'The temporal structure of organizations', in E.A. Tiryakian (ed.), *Sociological Theory, Values and Sociocultural Change*, London, Free Press.
Mumford, L. (1973), *Interpretations and Forecasts*, New York, Harcourt, Brace, Jovanovitch.
O'Connor, J. (1975), *The Fiscal Crisis of the State*, New York, St Martin's Press.
Ornstein, R.E. (1969), *On the Experience of Time*, Harmondsworth, Penguin.
Palloix, C. (1976), 'The labour process: from Fordism to neo-Fordism', *CSE Pamphlet No. 1*, London, Conference for Socialist Economics.
Piaget, J. (1969), *The Child's Conception of Time*, London, Routledge & Kegan Paul.
RSNT (Railway Staff National Tribunal) (1979), 'Decision No. 63. Reference to the RSNT of Claims by the Railway Unions for a Shorter Working Week', London, The Special Joint Committee on Machinery of Negotiation for Railway Staff.
RSNT (Railway Staff National Tribunal) (1982), 'Decision No. 77. Proposal by the British Railways Board for the Implementation of Sub-paragraph 2(c) of the 1981 Productivity Understanding to Footplate Staff', London, The Special Joint Committee on Machinery of Negotiation for Railway Staff.
Serpell, D. (1983), *Railway Finances*, Department of the Environment (Transport), London, HMSO.
Sève, L. (1978), *Man in Marxist Theory and the Psychology of Personality*, Hassocks, Harvester Press.
Simmel, G. (1971), 'The metropolis and mental life', [first published 1903] in *On Individuality and Social Forms: Selected Writings*, Chicago and London, Unversity of Chicago Press, pp. 334–9.
Sorokin, P.A. (1943), *Sociocultural Causality, Space and Time*, Durham, N.C., Duke Unversity Press.
Taylor, F.W. (1911), *The Principles of Scientific Management*, New York, Harper & Row.
Thompson, E.P. (1967), 'Time, work-discipline and industrial capitalism', *Past and Present*, vol. 38, pp. 56–97.
Vaneigem, R. (1967), *Traité de savoir-vivre a l'usage des jeunes genérations*, Paris, Gallimard.

Walliman, I. (1981), *Estrangement. Marx's conception of human nature and the division of labor*, Westport, London, Greenwood.

Weber, M. (1976), *The Protestant Ethic and the Spirit of Capitalism*, London, Allen & Unwin.

Willis, P.E. (1977), *Learning to Labour: how working class kids get working-class jobs*, Farnborough, Saxon House.

Zerubavel, E. (1981), *Hidden Rhythms: schedules and calendars in social life*, Chicago, Chicago University Press.

CHAPTER 6
'YOUR TIME OR MINE?' AN ANTHROPOLOGICAL VIEW OF THE TRAGIC TEMPORAL CONTRADICTIONS OF BIOMEDICAL PRACTICE

Ronald Frankenberg

It does not require a very introspective patient to see the surface connection between time and her experience of medicine. Medicine seems, like some of those described by the culture and personality school, to be a waiting culture. Surgeries or physician's offices usually have a waiting room and one waits to see the doctor a matter of minutes and hours, for treatments a matter of days or weeks and for 'cold' surgical operations a matter of months or years. Indeed a pharmacist employed by a major English drugstore chain testified to a Nuffield Foundation enquiry that while it was possible to dispense some medicines in a matter of seconds by reaching up and taking a package from the shelf, it was company policy to impose a delay so that patients would realise as a result of having to wait that drugs were not just commodities (and he might have added that pharmacists were professionals and not mere shop assistants).

For physicians, and other health workers as well, time is not peripheral but of the essence. The physician has first of all the longest period of training of any professional (save the veterinary surgeon) and during that training works longer and more irregular hours than any fellow student. He has the shortest vacations and during his postgraduate training his time is literally not his own. The notoriously and (in terms of patient care) irrationally long hours of what in UK are called junior hospital doctors and in the US interns can be seen as a symbolic initiation into their opposite, the sacred, inviolable time of the qualified physician which is protected and parcelled out by his supporters. (But see the paradox of the breach of sacred norms below.) There is the same asymmetry between the value of the consulting patients' time and that of their doctors as there is between the sovereign's or the president's and that of their attendant ministers and press corps and largely for the same reasons, both functional and symbolic. The principal actor in the performance is a scarce, even unique, good and has symbolic value above and beyond his scarcity value.

In Robert A. Hahn's fascinating but disturbing in-depth description of an internist (to English readers something between a general practitioner with hospital rights and a consultant physician) whom he calls Barry Siegler he writes (1985, pp. 61–2):

> Barry's life is scheduled far in advance. He knows when he will take his vacation next year, which weekends and nights he is 'on call', which mornings he will make rounds in which hospital, and the hourly contours of each day. These plans are generally changed only for emergencies – those of patients, colleagues, and his family. Patient emergencies are the business of being on call, but these and other emergencies may also interrupt the remainder of the schedule as well.
>
> Barry's time is structured by degrees of medical engagement – of access by patients, and of his reciprocal obligation to respond. Each partner has intense time 'on', but certain time 'off' as well. Each is 'covered' by his partners when he is 'off'. These states significantly define the physician's role. Barry once referred to periods spent in the hospital as 'clump time' and to periods at home waiting to be called as 'scattered time'. While in the hospital medical work is continuous; call at home leaves him continuously susceptible to medical interruption. Emergency call reaches the office [anglice surgery] as well. When I first went to meet Barry in his office, to explain my study and to ask to accompany him, he had left the office without warning to go to the hospital where his patient was admitted with an 'MI' (myocardial

infarction or heart attack). I left to return another time; scheduled patients were delayed.

Barry's schedule marks out his location and the general scope of his work there at different times, but not which patients he will see or what specific problems he will have to treat. The schedule provides slots for work whose size is limited in several dimensions. Certain problems can be dealt with in the hospital, but not in the office, or in the afternoon hospital working visits, but not in hospital rounds, or in the intensive care unit, but not 'on the floor', and so on. Barry tries to fit the work into his schedule. If parts do not fit, they are abandoned or postponed, or the schedule expands, extending, Barry's day, deferring and reordering other appointments. He 'runs behind'.

Barry is oppressed by two sorts of temporal constraints: too much to do in the allotted time, and the urgency of action imposed by what are too often literal 'deadlines'. Time is differently shaped accordingly: excess draws it out, imminence intensifies each moment.

Despite the gruelling schedule, in this case shared for the five months of the study by one of those academics whom Barry normally despises as having too much time on their hands (Hahn 1985, p. 78), the symbolic asymmetry with patients other than the actually dying remains (Hahn, 1985, pp. 69–70).

Barry always tries to leave the hospital for the office around 9 am, so that he does not make office patients wait. He rushes to leave, impatient; he often postpones until the afternoon what he planned or hoped to do in the morning. Still, he notes, patients sometimes wait as long as two hours to see him. Most patients are scheduled far in advance; others, 'squeeze-ins,' call with problems felt to be urgent, and are seen if the partners and/or staff concur on the urgency. I suspect that except in the most extraordinary of circumstances, perhaps a serious plea from a valued community colleague, only Barry's or his partners' own patients would have this immediate access to his attention. Barry says that he himself waits about five minutes every five years, when patients do not show up.

It is clear that the expenditure of time, like shifts in space – home, surgery (or in US office), hospital – has important symbolic value in the ambience of medicine and healing, what I choose to call the cultural performance of sickness. This, of course, has not hitherto passed unnoticed whether by the patients who suffer and perhaps thereby earn their cure through it, by the consultants who manipulate and enhance their status through it, or by sociologists like Zerubavel who have recorded it and thereby received their doctorates. Turner (1985, pp. 227–46) in a paper first published in 1981, introduces a concept of Anti-Temporality to analyse the immunity from sanctions of an irreverent, treasonable and satirical clown within a South Indian temple ceremony. In those circumstances his performance was seen by those present as taking place in 'timeless time'. One can contrast such absolute anti-temporality with the relative anti-temporality which distinguishes hospital from world outside. Time inside is real enough as is time outside. Each, however, has its own idiom independent of the other but transformable by the ceremonies of admission for patients, donning white coats and name tags for the staff. Thus the pre-operation abstentions are only regarded as reliable if the patient spent the previous night in hospital; the apparently healthy are moved from front door to ward by wheelchair and impertinent penetrations into the body physical are made as pertinent as the holy clown's penetration into the body social, not by necessity, but by translation into a new world of time. (The general point is made by Goffman, 1968.)

This chapter seeks to analyse the reality with a greater emphasis on the study of the culture, in addition to the organisation, of health, sickness and their associated institutions.

A CLASSICAL MEDICAL SOCIOLOGICAL APPROACH

From novels and personal reminiscences as well as such sociological studies as Julius Roth's *Timetables* or Zerubavel's

Patterns of Time in Hospital Life we can see that for the patients the passage of time in hospital appears to continue inflexibly from routine to routine with almost total disruption of the 'normal, natural' rhythms of bodily desires; to sleep or to go on sleeping, to eat or to refrain from eating, to evacuate waste solids and liquids, and to enjoy sexual experience. Indeed as the nature of jokes in hospital films and television series shows no time or no space at all is legitimately allowed for the latter. (For a poignant and suitably sanctioned exception see Collier and Horowitz, 1985, p. 540.) A whole genre of comedy is based on these disruptions, both iatrogenic and bureaucratic, of the temporal basis of motions across body boundaries. Even in 'Hospital Romances' it is only significant glances that are exchanged in hospital time. Penetrations of the mildest osculatory kind take place in off-duty times and places.

The time experience of both doctors and nurses is not only markedly different from that of patients but is also itself quite diverse.

For nurses there is a scheduling of time even more rigorous than for the patients since it is both externally sanctioned and ingrained. Within a variation imposed by the exigencies of a system of shifts, nurses come to work at fixed times and proceed through a liminal period of changeover, to a fixed period shift. When the shift ends, whatever crisis may be in train, it ends definitively for the nurse and her shiftmates. She goes off duty for a determined, bounded period of rest and does not, in any circumstance, return until she is next on duty. In Strauss's (1985) terms sentimental work may be an essential part of the nurse's duties for the dying and for the giving birth but the sentimental worker like other workers is not merely substitutable but has substitution imposed upon her by clock-regulated schedules. For nurses then, private time and public time are clearly demarcated; and so thereby is their use of the public space of the ward and their role in the lives of their patients. In US hospitals nurses' work time is measured and paid for by the hour (Zerubavel, 1979, p. 16). The continuous vigil at the bedside is reserved for kin, parents of immature sick children, husbands of those in labour, expectant widows and the mature children of sick and dying parents.

The yearly paid doctors, of course, also have their fixed points; ward rounds, clinics and fixed times for particular procedures when it is necessary that patient, nurse, technician, porter and physician or surgeon should all be in the same place at the same time (Giddens 1984, p. 114). However, there is no time of day or night when the expectation of their presence is unthinkable or when they can not, at least in theory, be summoned. Nor is the appearance of a doctor on the ward at any time, whether or not he/she is known to be on duty, in principle the matter of remark or speculation. The ultimate sign or perhaps even symbol of this is the bleep (US beeper) which has supplemented the label, the white coat and the stethoscope. While the latter emphasise the incumbency of a role in space and at a particular time, the bleep implies that the persona and the individual are never separated for long even in sleep, at meals, in the washroom or the erotic embrace. The boundaries of private and public space and time are, again in theory, as totally disrupted as they are for the inmate of the total institution. But whereas in the latter context this marks a loss of status, for the hospital doctor it is a symbol of almost sacred power and powers. For him there is not only no boundary between private and public, but also none between night and day and hence by extension between death and life. The fact that night time as a 'frontier' (Giddens 1984, p. 119) is so marked in the hospital, as every patient knows, serves to emphasise still further the mystery of those who cross it so easily. This, of course, is only fitting for those licensed in the course of duty to cross, even to penetrate, the boundaries between the embodied self and the self-in-the-world.

Zerubavel's conclusion that the flexibility of time scheduling is inversely related to status and power within the hospital, while it stimulated this analysis, is perhaps too simple to encompass the symbolic realities involved. Just as St Francis and his followers derive their holy powers (and my use of the plural is deliberate) not from being poor but from their choice of poverty, so senior doctors (e.g. in UK, consultants) derive theirs from the choice of forgoing the privilege of private time. The choice is not available to nurses and their status is diminished because they are denied it. The time status of hospital midwives, not specially summoned back to duty but allowed to remain to see a birth through, provides an interesting limiting and intermediate case.

While it seems to be the case that natural rhythms decree that deaths, disease crises and births most often occur at night, this also has social and cultural implications. It may be that the revolt of women against inductions and caesarian births was made possible precisely by the fact that shifting birth to the daytime in these ways disrupted the symmetry of the equation: I have violated the boundaries between the physician's public and private time, she or he may violate the boundaries of my body self. It now became apparent that the obstetrician was exercising power instead of powers and his authority became literally and metaphorically questionable. Those physicians who wish to abdicate the responsibility thrust upon them to turn off life support machines should ensure that the operation is carried out between nine and five. Italian primary care physicians who work similar office hours have similarly deprived themselves of both mystery and status which the Guardia Medica, who take their place at night, the weekend and on public holidays, do not gain since they are known not to be volunteers but pressed men (and women) compelled like any other night shift worker by economic or family necessity.

OPPOSING APPROACHES TO MEDICAL ANTHROPOLOGY

Arthur Kleinman in his seminal and important works which have done much to establish the bona fides of the subject amongst US physicians and especially psychiatrists makes a distinction between disease and illness. He transfers to the study of illth in western society a distinction spelled out and analysed in, for example societies like the Gnau in New Guinea, by writers like Gilbert Lewis (1975, pp. 148–9). In doing so, Kleinman is able, for the most part, to reassure his fellow physicians that their concepts of disease do not have to be rejected, their training is sound and their understanding of pathological entities good. However to be at their most effective in relation to curing or succouring their patients they need the understanding of the meaning of disease to the sufferer which the study by medical anthropologists can illuminate, analyse and in the most favourable circumstances, explain. This he calls illness (Kleinman, 1978 and 1985). Medical care then, again in Strauss's general terminology of arcs of work (1985), involves educational work. The physician must learn the patient's interpretation, the nature of the illness, and then in turn as part of the cure, teach the patient as much about the disease as is possible and necessary for treatment to be both carried out and complied with.

While such an approach certainly presents a conceptual advance and has led to others as well as much valuable empirical work, I have in several earlier publications argued my feelings about its limitations. The approach seems to me to beg questions about the nature of healing by privileging unduly both the physician's definition of the nature of the illth and his role in bringing about cure or providing care. In this chapter I hope by introducing ideas related to time to carry this criticism further. I feel this particularly apposite because in the context of academic study such an approach also tends, in my view, to make it more difficult to apply to western medicine that kind of critical sociological attention which we give to systems which are less readily acceptable to western scholars. At the same time it may, therefore, weaken our study programme in the sociology of science which holds that the sociological explanation for the acceptance of mathematics should not differ in kind from that needed to explain the acceptance of phrenology or ectoplasmic spiritualist manifestations (Bloor, 1976; Law and Lodge, 1984, p. 185).

My own preferred conceptualisation (Frankenberg and Leeson, 1976), derived originally from Suchman, but also having emic roots in the English language, is to add to Kleinman's disease and illness a concept of sickness which refers to the way that illth is socially and culturally performed. Paying attention to such a process involves the scholar in a fruitful questioning of a set of interactions. In the western case relationships will be revealed between individuals and things which include physicians, other health workers, sufferers, relatives and friends as well as diverse objects with both instrumental and expressive functions in the total performance.

By disease, I mean a disturbance of body functions and performance seen in biological terms of the kind which physicians are primarily trained to detect, diagnose and treat. Biology is also concerned with time both as the background to the evolution of life and its forms and, in smaller units, as a measure of the processes of morbidity and mortality. A disease, of course, can be seen in such terms as a temporary but potentially permanent interruption to the process which I have elsewhere called lifedeath in order to emphasise its finitude (Frankenberg, 1987), or as a speeding up or slowing down of life processes or as a cutting short of lifedeath itself. However, as Gabbay (1982) has cogently pointed out, it is a common illusion of biomedicine to assume that part of the definition of diseases as diseases is their timelessness or perduring nature, whereas, as he has shown for asthma, its legitimate definition even at one point of time is hotly disputed within and between treatises.

Illness is concerned with the conceptions of sufferers of that which physicians conceive of as disease. Sufferers may observe themselves as being unable to achieve against a measurement of time what they could achieve before their illness. The meaning of segments of time may be disrupted or destroyed. It is still 8 o'clock but no longer time for breakfast and times for play and for work may no longer have that subjective meaning. Enhanced consciousness or mortality may change time meaning even more fundamentally. Again, as Zola (1972) pointed out, neither patients nor their friends, relatives or doctors, hesitate to move the onset of their illness back in time beyond the incidence of their disease. They recall staying out in the rain, not wearing warm clothes, not being careful where they trod, eating, drinking and smoking inappropriately and in general neglecting to keep themselves healthy. They thereby date the origins of, and by implication responsibility for, their misfortunes in their own past misdemeanours. It may be that, for this and other reasons, in addition to distinguishing the best of bees from the worst of architects by their differential ability to plan the future, Marx would have been wise to compare also their capacities to reconstruct the past.

To me sickness is the key concept since it provides the

social and cultural framework for the anthropologist to analyse the social consequences of illness and disease which are in this sense secondary. Disease depends on the existence of the social organisation of biologists and the medically trained and illness on a socially constructed sense of self which certainly does not exist in the same form in all societies. Sickness as cultural performance can be analysed in all societies even if a specific performance may be the denial of disease and the absence of illness. The performance of sickness is always culturally specific not only to particular times and places but also in relation to conditions of disease occurrence like epidemic, endemic or individual. This is particularly clear amongst the Gnau, for example. They have a spirit, Taklei, who causes epidemics (Lewis, 1975, pp. 81–2 and pp. 168–9), they do not bother to explain endemic diseases and they account for individual sickness in a large number of different ways which Lewis shows are related to spirits of various kinds, to sorcery and magic or to breaking of taboos. Actual cases in each category are explained by past events related to time sequences, places, the ingestion of food and particular actions. Sickness amongst the Gnau as in all other societies always involves, by definition, a usage of time and space different from a situation where sickness is not performed. Cultural performance of sickness may or may not be influenced by changing views on the etiology of disease (Gabbay, 1982, p. 26); it is certainly and always revealing about social relations in general since it is concerned with the management of the body and its boundaries, the relationship of society to nature (Comaroff, 1982, p. 51) and with mortality and the transitions lifedeath, deathlife and death. I have introduced these concepts in an earlier paper (Frankenberg, 1987) seeking to operationalise Heidegger's concept of 'life-towards-death' (Baumann, 1978, p. 165). Most empirical sociologists who have taken an interest at all in such topics have concentrated on deathandying conceived as one word which, on the one hand, feeds the illusion that all but the final stages of the life course can reasonably be analysed without reference to its universal and inevitable end and, on the other, leaves the problem to hospice workers and specialist social scientists. Lifedeath is used in place of that part of life in which death is part of what Bloch (1986) calls the 'Not yet conscious'. Deathlife is used to indicate that part of life which is lived in the consciousness of its not too far distant end. Just as in the case of illness the cultural performance of sickness, like other performance, may require more or less systematic, notional reconstruction, through selective inattentions, of past actions through which present and future performance can be understood (Lewis, 1975, pp. 227–8). There is an analogy with the theatre where we usually meet characters fully mature and rely on their dialogue and non-verbal signs to deduce a past and a future in this case even more fictitious than their present. Roth has a striking example in his study of tuberculosis hospitals (Roth, 1963, p. 28)

> . . . when a medical resident presented a patient as having been in the hospital for two years, a consultant promptly announced: 'That's a long time; I think we should try and get her out of here.' One of the nurses pointed out that the resident had made a slight mistake on the matter of length of hospitalisation. The patient had first been admitted about two years ago, but had been discharged after about a year, had spent half a year on the outside, and had then been readmitted. Her second admission had involved slightly less than her six months' stay in the hospital. The consultant replied; 'I'd hate to let her go too soon this time.' This doctor then argued in favor of holding the patient for at least another conference six months later. All this time the physician was looking at exactly the same set of X-rays and was considering the same information concerning bacteriological tests and other diagnostic procedures. The only reason for the different position he took in his second statement was knowing that the patient had been out of hospital for a period of time and that her present period of hospitalization had been for only a relatively short time – obviously *too* short a time, in the doctor's opinion.

The cultural performance of sickness is an episode in not one but several time sequences for each of the performers. For a surgeon, a tonsillectomy may be merely the first of the day or the first of his career, but it is or is expected to be part of a series of repetitive events which make up his week, his year, his career and his life. For the patient it may mean a week removed from her life course and one that has been subjected to insertion in a different framework of time and space. For patients with burns or recurrent cancer or for leukaemic children (Bluebond-Langner, 1978) certain performances may be as familiar in a different way and as repetitive as for the surgeon. The whole life-course organisation of time for many *chronic* patients, as the descriptive adjective we use for them unwittingly implies, may be organised around dialysis or injections of insulin. Comaroff (1977) has described the battle joined between, amongst and inside attenders at clinics and their attendant midwives and doctors over whether about-to-be mothers are well or ill. This too can be seen in terms of time labels; pregnant for nine months, expectant mother embedded within a life course and in-labour for a period hopefully and usually measured by clock rather than calendar. (See Rosengren and DeVault 1963 for an analysis of the use of space, barriers and lighting to mark the boundaries of microperiods within the longer time-period of medicalised labour in a hospital setting.) The analogy but not the congruence with respectively disease, illness, and sickness is I think clear, as is the similar linguistic division between the chronic patient, the invalid or cripple, and the disabled.

This draws our attention to a point which Strauss et al. (1985) have made central to their analysis of the modern US hospital. The rarity therein of the acutely sick, combined with the high technology nature of diagnosis and treatment means that the knowledge and understanding over time of the sickness trajectory enjoyed by technicians, nurses and especially patients and their relatives may well be much greater than that of their physicians. In general the physician may know more than his patient about acute disease in practical as well as in theoretical terms. He has seen many discrete patients over a period of years, the chronic patient on the other hand has had the opportunity to study one patient over many years and continuously. A collection of such patients organised into a specific patient group may have much to teach not only about illness and sickness but also about disease as, at least in the UK and US, is now widely professionally recognised.

Albert Kushlick suggested the importance of this long ago (Kushlick et al., 1976) when in evidence to a royal commission he suggested that the carers for institutionalised and other incapables could be classified by the time spent with the sufferers. There were, for example, 4-minute-once-a-week people, 8-hour people and 16–24-hour people and, in his view, responsibility for life decisions for the patients and the validity of the advice given to royal commissions should be judged in direct proportion to the time spent. Since such people were respectively, relatives, nurses, and physicians in descending order, he was assumed to have made a joke in poor taste.

POWER, SPACE AND TIME

Kushlick's suggestion was so serious that it could only, like effective satire, be treated as wit. It draws attention to that creation and control of power through the medium of time which is the core of all cultural performances of sickness. It leads us towards the appreciation of the tragic, because inevitable, contradiction of so-called scientific medicine. Its application depends on the timescale surrounding illness. To practice the art of healing the physician, like the fieldwork-practising anthropologist, meets the sufferer in her own time as a coeval. To practise the science of medicine and analyse and treat the disease the physician distances himself in time from the patient and treats her as allochronic, in another time. This is made paradoxically possible for the physician by the removal of the patient to hospital, the place which he controls but where he is not necessarily always co-present in practice, although, as we have already seen, he is always present in theory. The time transformation is implied in the views on history of Hahn's internist, Barry (1985, p. 91):

> 'Listen to the patient', Barry repeats emphatically to the residents. The members of 'the company kept' [i.e. the symptoms] can most often be ascertained from the patient him or herself. 'Nine times out of ten the patient

will give you the diagnosis.' . . . [verbatim quotations from Barry omitted] Part of what Barry listens to is the patient's 'history'. 'Your history is so damned important . . . Take that history for what it's worth.' The 'history' is a chronology of the patient's symptoms, activities, and environment. The 'worth' of this chronology, transformed from the patient's to a medical 'history' is essentially diagnostic. The history is cast as an impersonal thing: it is 'that' history and 'your' (the physician's) history, and it is taken by the physician from the patient, medically transformed and medically used.

Again, 'listening to the patient' and 'taking the patient's history' have the appearance of conformity to contemporary concern with the whole person, the patient's interests, individuality, context, and so on. Again the appearance is deceptive. Barry redirects the sharply delimited information drawn from the patient to his physiological examination. 'Listening to the patient' and 'taking the patient's history' are not efforts to sympathetically comprehend the patient's life world, inner meanings, fears or desires, but rather to diagnose disease conceived of by criteria independent of their personal features.

Our language is more precise than we may give it credit for when we speak of physicians 'taking' rather than 'listening to' patients' histories. (Pauline Hunt has pointed out to me the similarity between this and industrial situations where workers' views on policy are sought, apparently listened to, and then either not acted upon or used in an alien management context.) This is not the least of the reasons why it is useful to draw attention to the helpfulness of Foucault's analysis of both psychiatry in particular and medicine in general and the relationship of both to the development of society based on both 'discipline' and industry. In particular he demonstrates that, 'The asylum and the hospital are not contingent institutional "contexts" of psychiatry or clinical medicine. They must be included in its definition.' (Cousins and Hussein, 1984, p. 76).

The cultural performance of sickness involves very specific

movements through space and over time and Foucault (1973) has demonstrated the changed social pattern of this for particular kinds of patient at differing historical periods just as Goffman (1968) has spelled out the implications of such movements for the individual moral careers of mental hospital and tuberculosis patients and Turner (1968, Turner and Turner 1978) for the interactive spiritual perceptions of individuals and collectivities in both African 'cults of affliction' and Marian pilgrimage.

In prehospital healing, the sick possess the power of compelling co-presence of significant others across normal time boundaries. Thus amongst the Gnau, seriously unwell men move themselves aside from the places where they would be expected to be at particular times and draw attention to their bodies by covering them with dust and thus by non-verbal departure from norms of space, time and body surface indicate that they are ill. In many cases that is that, but in others relatives travel from other villages to be with them at times when they would normally be absent about their own business or the patient is transported to other places where he would not normally be at that time. Sickness has been thus initiated and its performance, according to circumstance, may or may not be extended into the past by offering explanations of offences against spirits, being in taboo places at taboo times, causing offence to sorcerers, etc.; or into the future by the performance of rites and rituals in addition to the ritual quality of being sick itself to which Lewis calls attention (1975, pp. 140–1). He sees sickness as a rite of passage – in this case out of society altogether – which he contrasts with Turner's description of some Ndembu sickness performances as marking the passage to new status as members of cults of affliction and I.M. Lewis's description of possession by pathogenic spirits as the normal road to a shamanistic calling (Lewis, 1971).

The power to command the spatial and temporal co-presence of relatives is a common feature of sickness in societies as far apart as the Nyuswa-Zulu (Ngubane, 1977, p. 101), modern Italy (Frankenberg, 1985) and mediaeval Jewry whom the Talmud cautioned against too ready an acceptance of a call to the bedside lest those not sick be encouraged prematurely to enjoy the powers which sickness brings as well as those which it disrupts. It even operates in some hospital systems like that of Lusaka where in the 1960s it was reported that only cure was offered, care and nurture remaining the responsibility of kin (Boswell, 1965). In those societies which, unlike the Gnau, have full-time healers, these may control the disposition in space and the short periods of co-presence of summoned kin. It is only in industrial society, however, that the continuous sharing of time, day and night, by patients and their relatives is overridden for long or short periods by medical staff. Even there, as we have seen, it is restored by impending death or birth, or by belief in the rights and duties of the nuclear family in conjunction with the need for continuous care beyond the time resources of staff (as in modern UK paediatric and labour wards).

The control over where the patient is at what time and over who is co-present with her or him is well described in A.E. Ellis's fictionalised account of his experiences immediately after the Second World War in an alpine tuberculosis sanatorium in Haute Savoie (Ellis, 1958).

The major mechanism of this control, so vital to the maintenance of the power relations which first bring patients to the hospital and then make sure that they conform, is the control of patients' time. One of the most perceptive subjective accounts of this is, unsurprisingly, that of Oliver Sacks (1984) who, as he reminds us, was a doctor for 15 years, before becoming a patient, and a patient who from the point of view of hospital staff was unnecessarily and impertinently concerned with his own time and his control over it. After his account of the depersonalising admission procedures, 'One is no longer a person – one is now an inmate. One understands that this is protective, but it is quite dreadful too' (Sacks, 1984, p. 28). He records his delighted surprise at being addressed by his name by a nurse and thus rehumanised and then the less pleasant first experience of having a 'history' taken from him rather than by him. This produced the worries common to patients and self-reflexive students of E.H. Carr and Braudel about what might be relevant let alone salient. The relatively junior doctors who question

Sacks tell him he is to have a general anaesthetic for the necessary exploration next day and when he protested:

> They said, no, general anaesthesia was the rule in such cases, and besides (they smiled) the surgeon wouldn't want me talking or asking questions all through the operation!
>
> I wanted to pursue the point, but there was something in their tone and manner that made me desist. (Sacks, 1984, p. 29)

He decided to exercise what he still saw as peer rights and talk to Swan, his surgeon, the following morning. After arranging his affairs by phone and a good night's sleep, he awoke at five, and remembered that the operation was scheduled for nine thirty and his unaccustomed viewpoint on the time drama was resumed.

> I waited eagerly for Swan to come. Six o'clock, seven o'clock, eight o'clock . . . Wasn't he coming, I asked Sister, a formidable-looking woman in severe dark blue (the jolly staff-nurse of the night before had been in a striped uniform).
>
> 'Mr Swan will come when he pleases,' she said tartly. (Sacks, 1984, p. 30)

At 8.30 he gets his pre-med injections from a nurse and at 8.45 he rings to ask what he had been given. He reports that after the injection he had not taken his eyes off his watch. 'Mr Swan made his appearance at 8.53 and found me gazing at my watch' (Sacks, 1984, p. 30). The surgeon turned out to be hearty rather than helpful and the familiar 'How are we doing today?' and 'Nothing to worry about!' were more or less communicated on the run with other cliches. Sacks rang the bell again for the now, obviously impatient, sister.

> 'What is it?' she said. 'Why have you called me?'
>
> 'Mr Swan,' I said, enunciating my words carefully. 'He didn't stay very long. He was just in and out. He seemed to be in an awful hurry.'
>
> 'Well, I never,' Sister huffed. 'He's a very busy man. You're lucky he looked in at all' (Sacks, 1984, p. 31)

Dr Sacks continued to be concerned with the slipping away of his control over time through the characteristic counting away of consciousness imposed by the anaesthetist, his final conscious memory of complaining, apparently correctly, that the second-hand of the preparation room clock had stuck, to awakening to tell a houseman that a few minutes ago his knee had felt fine and discovering that seven hours had passed; and lastly to forty-eight hours of semi-consciousness. At one stage he does recall his experience as a new boy at school and compare himself to a prisoner. His knowledge and former status enables him, notwithstanding the linguistic shift from Swan to Mr Swan, in some sense to distance himself and thus to prolong resistance beyond that possible to those totally encompassed within the system.

Hahn's Barry Siegler had to turn would-be patients away despite the fact that they were paying, had the choice of other doctors and had sometimes, by his own account, to wait two to three hours in his office for their average 12.3 minutes. Presumably they knew that he worked through the day without lunch to fit them in and hoped, sometimes correctly, that his obsession with anti-coagulants and his general treatments would prolong their lives (Hahn, 1985, p. 70 and passim). The case study of Siegler reveals very vividly a time organisation of medicine which would be absurdly irrational if all that was at issue was a kind of patient care analogous either to other forms of concern for the individual or even to the care of mechanisms that have gone wrong. We are not surprised to learn that physicians' time is organised differently from either priests' or motor mechanics'. That does not necessarily make it ideal for healing people.

Roth (1977) describes in detail the performance undertaken by himself and his son in order to get the latter 'excused from games' at his English public school. This involved the use by the authorities of delay as a deterrent, numerous appointments always at the doctors' discretion and an insistence on transferring the boy to a place where his time was no longer his own. Rather surprisingly, given that he originated the topic of time in medical settings (Roth, 1963, and Roth and Douglas, 1983), Roth does not draw attention to these temporal aspects but to other ways in which medical power is exercised.

Roth's failure to utilise the insights about time that he has taught to the rest of us also has profound political implications. It is perhaps the reason why his remedies learned from some aspects of the women's movement, like Anselm Strauss's, are merely radical rather than revolutionary in that they offer hope but combine it with despair. Despairing of fundamental reform of the physician or formal service organisations, their hope comes through self-help organisations, consumer education and patients being accompanied by advocates. Revolutionary changes in health services, like those elsewhere, require that time itself is turned upside down. The message of Marx's *Capital* to workers, that they should take charge of their own time, whether it is interpreted in terms of state socialism or co-operation, applies also to patients.

Adults in industrial society outside of hospital do, albeit within a framework of publicly scheduled events (like opening hours, times of performance, transport timetables, programme times, etc.), organise their own time in their own place but they have it organised for them in places where they are constrained by work or public regulation. Such places of constraint include, of course the home as the locus of child care and care for the elderly and infirm. Classical (Jahoda et al., 1933, reprinted 1972) and recent studies of male unemployment suggest that uncertainty of time scheduling is, after an initial period of the enjoyment of time anarchy, the first major problem to be resolved. Removal of patients, whether male or female, to hospital or even to clinics, whether outpatient departments of hospital or consulting rooms of GPs, puts them in a space where the patients' uncertainty is reinforced by the certitudes of the physician's (and to a lesser extent other health workers') understanding of complicated routines. Paradoxically when the timing of events such as release from long-stay mental hospital, or in the past, tuberculosis sanatorium, is almost as uncertain for the staff as it is for the patient, this too becomes a means of social control through real or apparent powers to make it less so. This is a major theme in literature, of both Ellis's *The Rack* and Mann's *The Magic Mountain* as well as in sociology,

Julian Roth's *Timetables* and as part of more complex analyses in Goffman's *Asylums* and Davis and Horobin's conclusion to *Medical Encounters*. In all these works we read of patients' attempts to reduce the uncertain timing of future events by wresting from physicians the knowledge of probabilities or by inventing certainty and progress by devising ways of literally 'marking' time (Davis and Horobin, 1977, pp. 217–18). It is one of the reasons that militant nurses in the United States are claiming the right, at present held exclusively by physicians, to fix the date of patients' discharge.

We can then develop Kushlick's classification by contact time with patients in another way by looking at tense in the discourse of healing. As we saw in Sacks's anxieties about the salience of events in his personal history, entry into clinical control does not merely imply power over the sufferer's present but also of his past and as we have just seen his future. In Kushlick's terms the agent who is least present, the physician, has the power to re-present the past and the future. The nursing staff, whose presents are long but fragmented, normally control only segments of the now. The 24-hour people, whether patients themselves or their relatives, whilst their omnipresence theoretically gives them the power over time – past, present and future – are in practice rendered achronous by uncertainty. Being made to wait and not knowing what is to happen and when, is not an incidental of the cultural performance of sickness in biomedicine-dominated healing, it is central to the conversion of the suffering to the patient, what Zola (1973) called the pathway from the person to the patient. In the terms of Fabian (1983) to whose characterisation of anthropology I am about to turn, it is the way that medicine creates its other and makes its object. Conversely a medicine that saw other health workers, let alone the patients, as equal participants in the healing process would neither need nor be able to treat the time of others as in its control. If Strauss et al. (1985) are correct in their diagnosis of the nature of both disease and its effective treatment in modern urban industrial society, namely that the former is chronic and the latter, hitech, there is in fact no other rational choice. To achieve their goals, hospital physicians, at least, have to share decision-making not only with other health workers and technicians (present in the hospital or not) but also with patients and their significant others. The very recognition of this necessity would transform biomedical treatment into a more complex form of social healing as already happens in some hospices and paediatric units. I, however, share the more pessimistic/optimistic view of Marx that those who hold power do not necessarily relinquish it merely because they can no longer exercise it effectively. Historical changes, like women in labour, still need midwives, even if in both cases they can most usefully be chosen from amongst their friends.

THE TRAGIC CONTRADICTION: IN OUR TIME AND IN THEIRS

Johannes Fabian, in a stimulating and important book (1983) which forces us to regard anthropology in an entirely new light, draws attention to the significant peculiarity of the ethnographic present. Social and cultural anthropologists, unlike most other social scientists and all historians, usually or often live alongside and share the experiences of those they study and whose society or aspects of whose society they intend to analyse. They become coeval with them while they collect their material. The anthropological exercise, however, also involves the scholar in reliving the experience in his mind and re-presenting it in speech and in print for an audience other than those who were his subjects. Paradoxically, if we follow Fabian's argument, the anthropologist by presenting his material in the present tense freezes the past in a way that establishes the other-timeworldliness, the allochronicity, of those he has studied. The present tense, Fabian argues, is a signal identifying a discourse in an observer's language. It separates observer from observed; the observer has moved on, usually to another place as well as to another time. The observed remain crystallised in space and time. I tell my students, the Dinka *are* raided by the Nuer and absorbed into their lineages; unless they are unusually alert they do not realise I am talking about what Evans-Pritchard thought that he could deduce from his observations in 1930 and 1935–6 and finally sent to press in 1940 when I was 10 and my students' parents were perhaps as yet unborn. Neither my students nor I would normally, as anthropologists, make a connection between those political events in the British Empire (the bombing of Nuer herds and settlements) to which Evans-Pritchard does refer in the past tense and the realities of the Nuer past and their famine and civil-war-haunted present. Interestingly enough it is precisely when looking at Evans-Pritchard's classic analysis of 'Nuer Time-Reckoning' (1939) that one becomes most conscious of the effect of this way of thinking about the other. For here and in the chapter on the same topic which substantially repeats it in *The Nuer* (1940), Evans-Pritchard makes the classical anthropological error of comparing the informal behaviour of the other with the formal of the self. The Nuer, when Evans-Pritchard studied them, saw time in structural and 'ecological' terms and had no clocks. (In Masai tourist villages in Kenya in the 1980s the proprietors made sure that the 'primitives' on display hid their wristwatches and transistor radios during visiting times (Bruner and Kirshenblatt-Gimblett, 1985).) In Oxford, the clock and the history-book chronology were present but perhaps not omnipresent. In their spare time even Fellows of All Souls may also have found it less difficult to discuss time with Nuer than Evans-Pritchard supposed. Gellner pointed out long ago (1958, p. 183)

> . . . how very general considerations, which in the past have arisen from analysing the relation between thought and things as such, have entered anthropological theory in the misleading guise of something specially intimately connected with anthropological material.

It is no wonder that anthropologists bringing their discipline back home to western Europe and the United States feel sufficiently uncomfortable to write and confer frequently about the epistemological problems and to seek to study what exotic other they can find even in their own countries (see Jackson, 1987 and especially Ardener therein) or to study the near at hand and to present it as other. Such discussions have led to the advocacy of a more self-reflexive approach to the interpretation of fieldwork experience which, as Fabian argues, implies the re-presentation to ourselves and others of our own past. Furthermore, if this intersubjectivity is to extend beyond fellow anthropologists we must, he suggests, be able to share each other's past in order to be knowingly in each other's presence.

Anthropologists can perhaps all too readily empathise with physicians. For some at least of these latter are faced with the problem that in order to get to know their patients they must be co-present with them as people in so fundamental a way as to be described as coevals. Paul Atkinson has amusingly and in another sense tragically described how on a shorter timescale, clinical teachers present the past symptoms of their successfully treated patients in the present tense. When patients have the temerity somatically or verbally to deny what is being demonstrated, they are reminded that physical and social coevalness are not identical (Atkinson, 1981). However, it is always important to distinguish between physicians involved in primary care and physicians operating as specialists. An analogy may be drawn with the division in the UK's other senior profession between solicitors and barristers. The former, like the GP, knows the circumstances of the client/patient in all their temporal and spatial density. The latter, like the specialist, puts the client's case in the context of the allochronic law court. A primary health care physician, GP in Britain, ideally and sometimes in practice, does not so much take a history (past) as take a present of the patient's social, psychological being. (It is instructive to compare the view and practice of Hahn's anthropologically intuited Barry Siegler with Berger's (1967) poetically evoked Sassal, the English country doctor of *A Fortunate Man.*)

It has been suggested (Howie, 1984) that in practice it is at the moment nearest to achieving present (coeval) empathy with the patient that GPs prescribe. In scientific theory, however, what they should do at this point is to diagnose, that is to say to withdraw their attention from the patient's social being (and by implication becoming) and turn it towards the patient's body and the biological processes to which it is, more or less passively, subject. Hahn's Barry Siegler, half GP, half consultant physician, is very conscious of the contradiction and speaks of his patient's problem as a 'picture', a 'story', a 'thing', (Hahn, 1985, p. 79). Living past/present becomes biomedical notional present tense. 'You have been overdoing it . . . exposed to infection, etc. etc . . . You have caught, developed X . . but . . X is, it has these characteristics.'

Although Sassall does certainly get closer to his patients, or perhaps Berger is a romantic to Hahn's realist, in the end even he comes back to biology, the body and a named disease.

> The doctor in order to recognize the illness fully – I say fully because the recognition must be such as to indicate the specific treatment – must first recognize the patient as a person: but for the patient – provided that he trusts the doctor and that trust finally depends on the efficacy of his treatment – the doctor's recognition of his illness is a help because it separates and depersonalizes that illness. (Berger, 1967, p. 68)

and again:

> No, he is acknowledged as a good doctor because he meets the deep but unformulated expectation of the sick for a sense of fraternity. He recognizes them. Sometimes he fails – often because he has missed a critical opportunity and the patient's suppressed resentment becomes too hard to break through – but there is about him the constant will of a man trying to recognise. (Berger, 1967, pp. 70–71)

and finally:

> When patients are describing their conditions or worries to Sassall, instead of nodding his head or murmuring 'yes', he says again and again 'I know', 'I know'. He says it with genuine sympathy. Yet it is what he says whilst he is wanting to know more. He already knows what it is like to be this patient in a certain condition: but he does not yet know the full explanation of that condition, nor the extent of his own power. (Berger, 1967, p. 74)

If general medicine is concerned with the present tense, specialities present surprising and paradoxical tenses. Thus surgery speaks to the future 'You will be better when I have removed the growth, the limb.' The most dramatic example is cosmetic plastic surgery that changes the present by creating a future from which part of the evidence of the past has disappeared. Psychiatry, however, at least in so far as it is based on psychotherapy, makes the extension of the present into the future bearable by permitting the restructuring of the past. In fact Bloch suggests that the great betrayal of Freud and Jung arose from their neglect of daydreams, aspirations to a better future, in order to concentrate on night dreams, a distorted present-ation of an unhappy past. He marginally prefers Freud who, although he wrongly saw daydreams as mere stepping stones to those of the night (Bloch, 1986, p. 86), at least tries to overcome the effects of the past on the material lives of his patients whereas Jung, according to Bloch, merely tries to situate his patients' immaterial psyches permanently not only in their own pasts but also in the mythical past of all humankind (Bloch, 1986, vol. I, pp. 51–113).

Bloch is angered by this because he sees in it the denial of the, to him, most important of human emotions, namely hope, which he sees as pointing to the future, and since he is a materialist, to the future of the socially situated body. It is with this last that the medical usage of time, past, present and future, is ultimately concerned.

CROSSING FRONTIERS: TIME AND THE HOSPITALISED BODY

It is of course an absurd truism to state that when patients are moved to hospital they are accompanied by their bodies, but there is nevertheless a necessity if discussing medicine and time to emphasise this. In fact, in Goffman's celebrated discussion of the admission procedures to long-stay institutions (1968), which are in many ways similar to hospital procedure, he emphasises the stripping off of integuments including clothing and supposed surface dirt, in order that the outsider stripped of clothes and roles may be reclothed as the first stage to socialisation as an inmate. Second, Foucault has drawn attention to the discipline of the body exemplified in the Rule of St Benedict and other monastic codes (Foucault,

1979), the introduction and development of military drill and methods of mass production in industry. Both Goffman and Foucault draw attention not only to the spatial but also the temporal implications of these trends; thus inmates of total institutions like those modelled by Goffman are always constrained to rise, retire and eat at fixed times and monasteries, barracks and mass production factories impose a rigid timetable on the body movements of those permanently or temporarily confined within them. Building on, developing and sometimes criticising these ideas, Armstrong (1983), Giddens (1984) and Bryan S. Turner (1984) have severally been able theoretically to illuminate the relationships between power, social organisation and the use and perception of the body, as in an earlier period did Mary Douglas (1973) inspired by Mauss (1936).

What I want to argue here is that relatively short-term and therapeutic rather than custodial hospitalisation is concerned with the body neither in its 'normal' socially constructed form nor as a totally biological entity. It is nearer to the latter and, as I have argued, biomedical models of sickness performance demand that this should be so. Nevertheless, especially at moments of loss of or return to consciousness, or of material flow across or through body boundaries, the breach of cultural norm obstructs the acceptance of the body's 'purely' biological nature even within the bounds of hospital anti-temporality. If we accept Mary Douglas's view that the natural body and the social body reproduce and reinforce the reciprocal images of one another then the removal of the patient's body to the hospital implies more than its biological reshaping. In fact, the total cultural performance of bio-medical healing is concerned neither with social persons nor with natural physiological or pathological processes but precisely with transitions between the two. The hospital body and the embodied patient are seen as bounded and what has to be controlled are movements within and across those boundaries. In this control, time is the essential factor.

In private life patients may or may not have been 'regular' in their habits; hospital routine tries to ensure that they are. Cartwright (1964, p. 69) optimistically interprets this as '. . . the strangeness of a new environment is succeeded by the security of a known routine . . .'. We have seen something of Roth's view, argued also by many others, that notwithstanding the power of the organisation and its certain (because fixed) temporal and other routines, the outcome of hospital order is a negotiated one, but one in which the patient negotiates from a position of temporal disruption and uncertainty. Ailon Shiloh's study of hospital patients in Israel led her to conclude that a bipolar distribution of patients existed in the hospital that she studied; those who saw themselves as partners to the staff (equalitarian) and those who saw themselves as acted upon (hierarchical). Both groups of patients were

> . . . prompt to reply, were articulate and replete with details as to the nature and cause of their pre-hospitalized condition at the very time that they were hesitant, brief and vague as to the treatment and prognosis aspects of their condition within the hospital. (Shiloh, 1972)

I am not, of course, suggesting that this is deliberately organised to give power to staff. It is merely a side effect of the focus upon, and the social and temporal restructuring of, transitions across the boundaries of the body.

The day's first such transition, which differs from the others in that material does not pass, is that from sleep to consciousness. In British hospitals at least the day starts early and although Cartwright's study appeared over twenty years ago, a hospital recently portrayed on television where patients were asked when they would like to be woken was presented as an exception. Of the patients in Cartwright's sample 95% were woken before 7 a.m., 62% before 6 and 35% before 5.30. Although many patients saw the nurses' point of view and none mentioned being awoken early directly as a reason for feeling critical of nurses, patients who were awakened early were in fact those most critical of nurses (Cartwright, 1964, pp. 38–9).

The Royal Commission report of 1979 also drew critical attention to unnecessary early waking, attributing it to tradition, the 'Nightingale' ward design and putting the convenience of the staff before that of the patients. (It did not draw attention to Cartwright's finding that the situation of teaching hospitals was better, probably because not only were there more nurses but they also worked a three- instead of a two-shift system (Cartwright, 1964, p. 171).) The Royal Commission explains the 'necessity' by the timing of shift changeover, itself dependent on the availability of public transport; medicine rounds; and the need to collect laboratory specimens. (They might have mentioned also the necessity of providing the patients with regular meals at relatively fixed times with limited staff.) All except the first of these involve a departure from 'normal' patterns of crossings of body apertures and translating of body fluids from outside to inside as well as extrabody substances in the other direction. Fluids leave at the wrong time, the wrong place and in the presence of the wrong significant others. In the Nightingale wards the specifics of shame are hidden by the screens around the bed; the generalities, by the hospital's time out of life. The rigid time structures of the hospital emphasise the anti-temporality of the experience in relation to 'normal' worldly time, just in the same way as the formality and timed ritual of the pilgrimage at Lough Dergh, for example (Turner and Turner, 1978, Ch. III). Thomas Mann's *The Magic Mountain* presents a set of variations on this theme for the extreme case of the long-stay tuberculosis sanatorium. Senior physicians, however, are less protected since, as we have already seen, they are at once privileged and condemned to breach both sacred and profane time boundaries as well as licensed and required to penetrate those of the body. Such a view helps us to understand the contradictions in Barry Siegler's combination of compassion and sexist vulgarity expressed in his own words and reported by Hahn (1985, pp. 103–6).

Hahn argues that two idiomatic strands pervade the way Barry talks about his work; an image of the body and of the dynamics of flow and obstruction within it. His movements about the hospital he sees in a similar way 'Barry is perpetually concerned with the qualities of channels of passage, with obstructions to passage, and with rates and regularities of flow' (Hahn, 1985, p. 104). These are charted for the patient who is visualised as a ship at sea. His favourite medicament, as we have already noted, is the anticoagulant and he is also vitally concerned with his own unobstructed flow through hospital corridors and overcrowded streets to keep the appointments on his overcrowded schedule. A similar imagery is applied to his efforts and treatment departments of this hospital and in and out of the hospital itself. It is, however, when things go wrong or are obstructed that Barry turns to scatological and aggressive, sadistic, sexual metaphor, unthinkable for Berger's Sassall or in extra-hospital practice generally. Patients and others are seen as turds or shits and Barry 'may tangle ass holes with them', (Hahn, 1985, p. 105). Things he disagrees with or is ambivalent about 'need to be screwed' and he makes full use of the love/hate ambiguity of cant terms for the sexual act. As Hahn gently comments

> Engagement with his patients' bodies touches him in a personal, emotional way. Beyond his medical concern for proper flow within the body, Barry is troubled by body effluence and excretion. Here the concern is one of contamination – contamination not in a medical sense, but in a more generic sense of dirt and impurity. (Hahn, 1985, pp. 105–6)

Hahn has earlier described Barry's moral disapproval and contempt – 'If you give a woman a vaginal suppository, you've got to tell her to take the foil off' (Hahn, 1985, p. 103) – of those he thinks of as promiscuous, the alcoholic and those who make their diabetes, for example, worse by their eating habits. In other words those who do not suitably control their body boundaries. Unsurprisingly, given what Hahn has told us about his religious background and beliefs, this disapproval is extended especially to women who, in the Christian as well as the Jewish tradition in which Barry has been socialised, are a particular threat to the clear maintenance of the timing of cultural natural boundaries. Both Jewish and Catholic concern with sexual intercourse and time of the menstrual month testifies to this concern. His attitudes to young women (whether patients or nurses; there is no direct evidence of his attitude to young women doctors) is in sharp contrast to his attitude to his favourite patients, no longer sexually or boundary-threatening 'little old ladies'. Apart from any personal idiosyncrasy of Hahn's subject, I would wish to suggest that the scheduling and lack of boundaries of Barry's time means that he is not protected as are other members of the hospital staff from the continual culturally uncontrolled natural breaches of body boundaries to which they are all subjected. At the same time he, alongside other physicians, has the power, the time, the captive subordinate audience and the licensed places in which to spell out his sadistic fantasies. Not all physicians indulge their opportunities as does Barry, but in UK and USA most have the same structural milieu. The nature of the jokes in the US-produced *M.A.S.H.* and the popular UK medical TV comedies and film suggest a rich vein for cultural anthropological mining.

CONCLUSION

In this paper I have merely begun to explore the interrelations of biological and social time in the cultural performance of sickness, especially when healing is based on a biomedical model. In some ways this is in fact the most interesting model to study since biomedical physicians and other health workers share an ideology of curing a disease: that is restoring or reconstituting a biological status quo. The stark choice of death or restoration to health leaves them little room for utopian visions (Bloch 1986, II, p. 467) or social fantasies such as might be indulged by benevolent African traditional healers cleansing a village of sickness-causing witchcraft or sorcery or even malevolent Californian politicians seeking to rid the world of sin, homosexuality, AIDS and their own unacknowledged, repressed temptations all at one stroke. I have found in the scholarly literature only hints of the different time worlds of old and young, male and female, rich and poor, dying and recovering patients and their similarly stratified doctors, nurses and therapists. I have had to have recourse, without great reluctance, to the fictionalised or 'unscientific' accounts of sick doctors and recovered sociologists and novelists.

Nevertheless it seems to me clear that this particular approach is revealing. In the mainstream of societies like the UK and US, the cultural performance of sickness within a biomedical framework takes place in a context where social, and especially temporal, mechanisms to control relationships of nature and culture, are disrupted. In order to maintain social order and restore natural order, patients are removed from their normal temporalities to a space where the time view of others can be imposed upon them. This is not incidental to, but an integral part of, the modus operandi of biomedicine as at present practised, hence my suggestion of tragic inevitability. It iatrogenetically produces a situation of enhanced power for the healers and reduced autonomy for the patients, notwithstanding the latter's attempts to negotiate their own reconstituted temporalities. As Oakley (1979 and 1980) and other writers as well as the recent experience of Wendy Savage in Britain have shown (Savage, 1986), and as we may well have experienced as parents, unsurprisingly if my analysis is correct, the contradiction of temporalities and of aspects of body boundaries is sharpest between social and natural motherhood on the one side and class- and gender-constructed obstetric medicine on the other. The designation of pregnancy as pregnancy rather than expectant motherhood, and also as disease, is a move in a power struggle in which the conceptualising of women's and of health workers' time is of central importance. As every English schoolboy used to know, the fact that authority may sincerely believe itself to be acting in the interest of the subject does not detract from its exercise of power or from the stimulus to opposition.

But in western medicine in general, time is more important as the basis of power and control than merely as the basis of value (important though Ricardo and Marx showed that to be in general). The problems that this creates for both patients and healers are not necessarily different in the totally capitalistic medicine of the USA, the partially socialised medicine of the UK, or the totally socialised medicine of the USSR. The more that China turns to bureaucratised, urban, curative, hospital biomedicine, the sharper will problems there become. Where diverse systems do differ is both in their capacity to resolve such issues without the total rejection of biomedical practice and in the relevance for them of its creating, maintaining and reinforcing the central ideological tenets of the society. These last are the two major overriding problems that require solution through social practice rather than mere research or scholarly presentation.

ACKNOWLEDGEMENTS

I am grateful to Michael Young and participants in the 1986 ASSET conference at Dartington for encouragement and constructive criticism of a preliminary presentation. I am especially grateful to Pauline Hunt and Ruth Frankenberg who read the whole of an earlier draft and made many helpful suggestions. My other debts are all too evident in the text.

REFERENCES

Ardener, E. (1987), 'Remote areas: some theoretical considerations', in A. Jackson, *Anthropology at Home*, Association of Social Anthropologists Monograph 25, London, Tavistock Press.

Armstrong, D. (1983), *Political Anatomy of the Body: Medical Knowledge in Britain in the Twentieth Century*, Cambridge, Cambridge University Press.

Atkinson, P. (1981), 'Time and cool patients' in P. Atkinson and C. Heath, *Medical Work: Realities and Routines*, Farnborough, Gower, pp. 41–54.

Atkinson, P. and Heath, C. (1981), *Medical Work: Realities and Routines*, Farnborough, Gower.

Baumann, Z. (1978), *Hermeneutics and Social Science: Approaches to Understanding*, London, Hutchinson University Library.

Berger, J. (1967), *A Fortunate Man: The Story of a Country Doctor*, London, Allen Lane, The Penguin Press.

Bloch, E. (1986), *The Principle of Hope*, Volumes I, II, III, Oxford, Blackwell.

Bloor, D.C. (1976), *Knowledge and Social Imagery*, London, Routledge & Kegan Paul.

Bluebond-Langner, M. (1978), *The Private Worlds of Dying Children*, Princeton, N.J., Princeton University Press.

Boswell, D.M. (1965), *Escorts of Hospital Patients: A Preliminary Report on a Social Survey Undertaken at Lusaka Central Hospital from July to August 1964*, Lusaka, Zambia,

Rhodes Livingstone Communication 29.
Bruner, E. and Kirshenblatt-Gimblett, B. (1985), 'Tourist performance and the ethnographic original', Annual Conference Paper of *American Anthropological Association*, Washington, D.C.
Bryman, A., Bytheway, B., Allatt, P., and Keil, T. (eds) (1987), *Rethinking the Life Cycle, Explorations in Sociology*, London, Macmillan.
Cartwright, A. (1964), *Human Relations and Hospital Care*, London, Routledge & Kegan Paul.
Collier, P. and Horowitz, D. (1985), *The Kennedys: An American Drama*, London, Pan Books.
Comaroff, J. (1977), 'Conflicting paradigms of pregnancy: managing ambiguity in ante-natal encounters', in A Davis and G. Horobin (eds), *Medical Encounters: The Experience of Illness and Treatment*, London, Croom Helm, pp. 115–34.
Comaroff, J. (1982), 'Medicine: symbol and ideology', in P. Wright, and A. Treacher (eds), *The Problem of Medical Knowledge: Examining the Social Construction of Medicine*, Edinburgh, Edinburgh University Press, pp. 49–68.
Cousins, M. and Hussein, A. (1984), *Michel Foucault*, London, Macmillan.
Davis, A. and Horobin, G. (eds) (1977), *Medical Encounters: The Experience of Illness and Treatment*, London, Croom Helm.
Douglas, M. (1973), *Natural Symbols*, Harmondsworth, Penguin Books.
Ellis, A.E. (1958), *The Rack*, Harmondsworth, Penguin Books.
Evans-Pritchard, E.E. (1939), 'Nuer time reckoning', *Africa*, vol.12, no. 2 (April), pp. 189–216.
Evans-Pritchard, E.E. (1940), *The Nuer*, Oxford, Clarendon Press.
Fabian, J. (1983) *Time and the Other*, New York, Columbia University Press.
Foucault, M. (1973), *The Birth of the Clinic: An Archaeology of Medical Perception*, New York, Pantheon Press.
Foucault, M. (1979), *Discipline and Punish*, Harmondsworth, Penguin Books.
Frankenberg, R. (1987), 'Life: Cycle, trajectory or pilgrimage? A social production approach to Marxism, metaphor and mortality', in A. Bryman, B. Bytheway, P. Allatt and T. Keil (eds), *Rethinking the Life Cycle, Explorations in Sociology*, London, Macmillan.
Frankenberg, R. (1974), 'Functionalism and after? Theory and

developments in social science applied to the health field', *International Journal of Health Services*, vol. 4, no. 3, pp. 411--27. [A longer version of this article is published in *International Journal of Health Services*, 1987.]

Frankenberg, R. (1985), 'Malattia come Festa: sickness as celebration and socialisation. Children's accounts of sickness episodes in a Tuscan community', Conference paper, Keele University.

Frankenberg, R. and Leeson, J. (1976), 'Disease, illness and sickness: aspects of the choice of healer in a Lusaka suburb', in J. Loudon (ed.), *Social Anthropology and Medicine*, ASA Monograph 13, London and New York, Academic Press, pp. 223–58.

Freidson, E. (ed.) (1963), *The Hospital in Modern Society*, New York, The Free Press.

Freidson, E. and Lorber, J. (eds) (1972), *Medical Men and their Work: A Sociological Reader*, Chicago and New York, Aldine/Atherton.

Gabbay, J. (1982), 'Asthma attacked? Tactics for the reconstruction of a disease concept', in P. Wright and A. Treacher (eds), *The Problem of Medical Knowledge: Examining the Social Construction of Medicine*, Edinburgh, Edinburgh University Press, pp. 23–48.

Gellner, E. (1958), 'Time and theory in social anthropology', *Mind*, vol. 67, pp. 183–202.

Giddens, A. (1984), *The Constitution of Society*, Cambridge, Polity Press.

Goffman, E. (1968), *Asylums: Essays on the Social Situation of Mental Patients and other Inmates*, Harmondsworth, Penguin Books.

Hahn, R.A. (1985), 'A world of internal medicine: portrait of an internist' in R.A. Hahn and A.D. Gaines (eds), *Physicians of Western Medicine: Anthropological Approaches to Theory and Practice*, Dordrecht, Reidel, pp. 51–111.

Hahn, R.A. and Gaines, A.D. (eds) (1985), *Physicians of Western Medicine: Anthropological Approaches to Theory and Practice*, Dordrecht, Reidel.

Howie, J.G.R. (1984), 'Research in general practice: pursuit of knowledge or defence of wisdom', *British Medical Journal*, vol. 289 (II), pp. 1770–2.

Jackson, A. (1987), *Anthropology at Home*, Association of Social Anthropologists Monograph 25, London, Tavistock Press.

Jahoda, M., Lazarsfeld, P.F., and Zeisel, H. (1972), *Marienthal, The Sociology of an Unemployed Community*, London, Tavistock Publications [first published 1933].

Kleinman, A. (1978), 'Concepts and a model for the comparison of medical systems as cultural systems', *Social Science and Medicine*, vol. 12, April, pp. 85–93.

Kleinman, A. (1985), 'Medical anthropology', in A. Kuper and J. Kuper, *The Social Science Encyclopaedia*, London, Routledge & Kegan Paul, pp. 511–12.

Kuper, A. and Kuper, J. (1985), *The Social Science Encyclopaedia*, London, Routledge & Kegan Paul.

Kushlick, A., Felce, D., Palmer, J. and Smith, J. (1976), 'Evidence to the Committee of Inquiry into Mental Handicap Nursing and Care' (The Jay Committee), HCERT Research Report 129, Winchester, Wessex Regional Health Authority.

Law, J. and Lodge, P. (1984), *Science for Social Scientists*, London, Macmillan.

Lewis, G. (1975), *Knowledge of Illness in a Sepik Society: A Study of the Gnau, New Guinea*, London, The Athlone Press.

Lewis, I.M. (1971), *Ecstatic Religion*, Harmondsworth, Penguin Books.

Loudon, J. (1976), *Social Anthropology and Medicine*, ASA Monograph 13, London and New York, Academic Press.

Mann, T. (1960), *The Magic Mountain*, Harmondsworth, Penguin Books [first published 1924].

Mauss, M. (1936), 'Les techniques du corps', *Journale de la Psychologie*, vol. 32, partially reprinted in T. Polhemus (ed.) (1978), *Social Aspects of the Human Body*, Harmondsworth, Penguin Books, pp. 35–6 from English translation by B. Brewster, *Economy and Society*, vol. 2 (1) (February 1973), pp. 70–88.

Ngubane, H. (1977), *Body and Mind in Zulu Medicine: An Ethnography of Health and Disease in Nyuswa-Zulu Thought and Practice*, London, Academic Press.

Oakley, A. (1979), *Becoming a Mother*, Oxford, Martin Robertson.

Oakley, A. (1980), *Women Confined: Towards a Sociology of Childbirth*, Oxford, Martin Robertson.

Rosengren, W.R. and DeVault, S. (1963), 'The sociology of time and space in an obstetrical hospital', in E. Freidson (ed.), *The Hospital in Modern Society*, New York, The Free Press, pp. 226–92.

Roth, J.A. (1963), *Timetables: Structuring the Passage of Time*

in Hospital Treatment and Other Careers, New York, Bobbs Merrill.
Roth, J.A. (1977), 'A Yank in the NHS', in A. Davis and G. Horobin (eds), *Medical Encounters: The Experience of Illness and Treatment*, London, Croom Helm, pp. 191–205.
Roth, J.A. and Douglas, D.J. (1983), *No Appointment Necessary*, New York, Irvington.
Royal Commission on the National Health Service (1979), *The Merrison Report*, Cmnd 7615, London, HMSO.
Sacks, O. (1984), *A Leg to Stand on*, London, Picador.
Savage, W. (1986), *A Savage Enquiry: Who Controls Childbirth?* London, Virago.
Shiloh, A. (1972), 'Equalitarian and hierarchical patients: an investigation among Hadassah hospital patients', in E. Freidson and J. Lorber (eds), *Medical Men and their Work: A Sociological Reader*, Chicago and New York, Aldine/Atherton, pp. 249–66.
Strauss, A. et al. (1985), *Social Organization of Medical Work*, Chicago, University of Chicago Press.
Turner, B.S. (1984), *The Body and Society*, Oxford, Blackwell.
Turner, E.L.B. (ed.) (1985), *On the Edge of the Bush: Anthropology as Experience*, Tucson, University of Arizona Press.
Turner, V.W. (1968), *The Drums of Affliction: A Study of Religious Processes among the Ndembu of Zambia*, London, International African Institute, Hutchinson.
Turner, V.W. and Turner, E.L.B. (1978), *Image and Pilgrimage in Christian Culture*, Oxford, Blackwell.
Turner, V.W. (1985), 'Images of anti-temporality: an essay in the anthropology of experience' [first published in *Harvard Theological Review*, 1981, vol. 75, no. 2, pp. 243–265] in E.L.B. Turner (ed.), *On the Edge of the Bush: Anthropology as Experience*, Tucson, University of Arizona Press.
Wright, P. and Treacher, A. (1982), *The Problem of Medical Knowledge: Examining the Social Construction of Medicine*, Edinburgh, Edinburgh University Press.
Zerubavel, E. (1979), *Patterns of Time in Hospital Life: A Sociological Perspective*, Chicago, Chicago University Press.
Zola, I.K. (1972), 'Medicine as an institution of social control; the medicalizing of society', *Sociological Review*, vol. 20, no. 4, pp. 487–504.
Zola, I.K. (1973), 'Pathways to the doctor – from person to patient', *Social Science and Medicine*, vol. 7, pp. 677–89.

CHAPTER 7

TIME: A SOCIAL PSYCHOLOGICAL PERSPECTIVE

*Marie Jahoda**

Every human being acquires in the course of growing up an intuitive grasp of that great ordering principle: time. But throughout the ages philosophers have struggled with the enigma of time and have not been able to agree on a way to think about this commonplace matter. Even a cursory glance at the philosophical literature makes one sympathise with St Augustine's despairing outcry in his discussion of time: 'For so it is, oh Lord, my God, I measure it but what it is that I measure I do not know' – at the very least a first indication of the difficulty in dealing with the topic of this paper, if not quite an argument against attempting it.

A second difficulty lies in the nature of psychology, including social psychology. Psychology is not a unified discipline. It ranges from natural science to social science to humanistic enquiries. Productivity under each of these differing approaches is so enormous that probably nobody can claim any longer to know it all; certainly not I. Even limiting the topic to studies of human beings in their natural social context – as I propose to do – and thus excluding the vast majority of psychological research leaves a very large field from which to gather approaches in the treatment of time. Since comprehensive coverage is out of the question I shall concentrate on only three issues that illuminate the treatment

* I am grateful to Prof. Lotte Bailyn and Drs David Fryer and Gün Semin for critical and constructive comments on an earlier draft.

of time. In the first of these the experience of time itself is the central focus. But time is implicated also when psychologists concentrate on changes in other psychological phenomena, where time is not the focus but the essential variable; this is the second issue. In both these areas substantial contributions have been made to the understanding of time. But time-conscious psychologists have raised a third issue in thinking about these achievements: will the insights so gained stand the test of historical time? To deal with this question requires a shift in perspective. The question is no longer how time is treated but rather how time treats the discipline. I shall deal with these issues in an inevitably selective manner by highlighting only a few relevant contributions.

Concern with time has waxed and waned throughout the relatively short history of psychology, a discipline that often ignores its long past in philosophy. Before psychology's unilateral declaration of independence philosophical psychologists had thought about time experience and formulated some insights that are as valid now as they were then. To give but one example William James (1890), one of the great psychologists of the past, formulated a universal aspect of time experience:

> In general, a time filled with varied and interesting experiences, seems short in passing, but long as we look back. On the other hand, a tract of time empty of experiences seems long in passing, but in retrospect short. (James, 1890, p. 624)

A currently unemployed man, unacquainted with James, echoed this formulation when he said that the hours of the day seemed almost unbearably long, but looking back the months seemed to have gone by in a flash. The relative independence of time experience from measured time is also expressed in one of the most frequently noted complaints of shopfloor workers engaged in repetitive tasks: time passes slowly. Such statements are invariably based on a comparison between experience and measured time, as is James's formulation.

It took an outstanding contemporary psychologist to attack systematically the experience of time and to put it to use in industry. Elliott Jaques's ideas on time were originally developed when he acted as a consultant on personnel assessment to a factory whose culture he had described in a well-known book (Jaques, 1951). Realising the importance of individual differences in time experience he emerged with the concept of 'time-span of responsibility' (Jaques, 1956), that is the longest [measured] period during which a person can implement an enduring intention [the psychological dimension] to produce successive events. For his practical purposes he used the lengths of these individual time spans as indicators of personal capacity.

Recently Jaques's longstanding involvement with the concept of time has culminated in a book (Jaques, 1982) that develops from an analysis of the philosophical controversies about time to a psychological approach with far reaching implications. He emphasises that time is a concept, not a thing that can 'pass' or 'fly'. In everyday language such reification is seductive because of mankind's marvellous inventiveness of instruments with which time can be measured, from microseconds to the age of the universe. Jaques's concept has two dimensions between which experiences oscillate: the experience of succession, of earlier or later, before and after; and the experience of the enduring present, the continuity of existence.

In this continuity the past is contained in memory, the future through desires, goals and anticipations. Past, present and future have conscious and unconscious content. He describes the enduring dimension as 'intentional', as a field force for action. This intentional dimension is conceptually distinct from the dimension of succession, but experientially they coexist. In the enduring Now, desires, goals and plans anticipate successive events and herein lies their measurability. Clock or calendar time can be attached to these anticipations as they can to the reconstruction of the past in the present.

The dimension of succession describes discrete experiences – one thing after another – as if changes occurred discontinuously. These snapshots through experiences are, as it were, at right angles to the experience of a permanent present. Jaques suggests that while the natural sciences since Einstein require four dimensions for the study of certain phenomena, psychology for some of its problems requires five, three for location and two for time.

These elaborated ideas help to interpret several independent empirical findings: the experience of time slowing down in monotonous work that leaves little room for subjective intentions; and the disturbances in the time experience of the unemployed where even the imposed intentionality of actions is missing and where the succession of events during the waking day has been largely eliminated; and the fact that a small minority of unemployed people maintain their time experience intact. It is particularly research on unemployment which is illuminated in this manner.

During the great Depression of the 1930s the study of Marienthal (Jahoda, Lazarsfeld and Zeisel, 1933), an Austrian village in which virtually everybody was unemployed, shows as a major effect the destruction of people's time perspective. Their measured walking speed slowed down. When asked how they had spent the morning, they named an activity that could not have lasted longer than a few minutes. Men stood around at street corners, but their wives complained that they had become unpunctual in returning for meals. Time had lost its meaning; there was nothing to look forward to, no hope and no planning for the future. When time had been structured through work, leisure had been fully used in a variety of voluntary organisations; unlimited time, not structured by successive events was experienced not as leisure but as emptiness. Modern studies of unemployment confirm that a major psychological consequence of unemployment is its interference with subjective time experience.

There are exceptions from this demoralising loss of time perspective now as there were then: some unemployed do not so suffer (Fryer and Payne, 1984). They experience the imposed independence from regulation by the clock as an opportunity to construct their own time structure. Majority and exceptions alike illuminate the essence of time experience: it is a function of intentions carried out. For the majority these realised intentions are restricted to minimal survival activities – getting up, eating, sleeping; for the exceptions they are extended to self-chosen activities of longer duration. The vast majority as well as the exceptionally active unemployed highlight the simultaneity of both dimensions in normal time experience. Without successive events enduring intentions shrink into insignificance.

Jaques's approach is still relatively unused by others; at least there are few, if any, direct references to him. Yet there are in psychological research some trends which, on a less abstract level, follow similar lines of thought. I refer to the increasing interest in studying time experiences in relatively enduring personal relations, such as friendships or marriage, or studies of the ability to delay gratification and its broader formulation in the studies on the length of time horizons, both with their well-known social correlates on which Helga Novotny (1975) has extensively commented. They could be interpreted within Jaques's scheme and so show the way toward some integration of the various approaches to the study of time experience.

It is more difficult to see how the contributions to this issue by the two greatest psychologists of the century, Piaget and Freud, could similarly be so integrated, perhaps because of their own fully elaborated theoretical ideas.

Piaget (Gruber and Vorrèche, 1977) reports that it was Einstein whose question to him at the 1930 conference stimulated his studies of time. He asked himself whether the intuitive grasp of time is primitive or derivative and set about searching for an answer in his characteristically inventive observational way. For example, he had children observe two toy trains, starting and stopping at the same time, but one covering a greater distance. Up to the age of about five, children tend to say that the train that covered the shorter distance stopped earlier. Spatial and temporal experiences are confounded. At that age children also fail to grasp the relation of age to birth order and confuse age with height. Spatial positions are available to immediate perceptions, while the concept of time requires an abstraction from the here and now of the visually given. Since children can alter positions in space at will they seem to make the assumption that the succession of events can also be reversed. I heard a fair-minded three-year-old object to being sent to bed earlier than his older brother, but comfort himself with the statement: one day I will be older and he will have to go to bed first.

There is in this remark evidence that one dimension of the time experience – the anticipation of the future in the present

– comes earlier than the train experiments suggest. But Piaget's exclusive concern with the acquisition of logical cognitive competence made him insensitive to this aspect of time experience, though he himself had shown even with pre-verbal children that the past – a hidden toy that is available to present consciousness and leads to appropriate goal-directed action in rediscovering it – continues in these very young minds.

Freud, on the other hand, as part of his concern with many aspects of time, illuminated time experience particularly in the dimension of the enduring Now in which past and future are present in conscious or unconscious memory, desires and intentions. When he described the unconscious as timeless, he referred to its freedom from the tyranny of clock and calendar, from the unalterable succession of events. Therapeutically and theoretically he elucidated the power of the past in the present (an aspect, incidentally, which Jaques does not elaborate). That early events may have repercussions in the present, for example anal fixation, has been ridiculed by many psychologists. Freud was, of course, talking about the past as living in the present, not of events that were occurring then but left no trace in the mind. The past can be in the present only to the extent that at the moment of the original occurrence it made an impact. Without this assumption the belief in the power of the past would be on a level with the primitive teleological belief that the future which by definition has not yet occurred can affect the present.

Freud, even more than Jaques, emphasised the continuity of existence from birth to death, the enduring Now of the 'timeless' unconscious, at the expense of underemphasizing the more reality-tied dimension of succession. From the perspective of social psychology it is the oscillation between the two dimensions which best captures normal time experience. In the modern world where the successive dimension is so often externally imposed, not internally planned, through the institution of employment, time experiences are disturbed in many when this institutionalised support fails.

Time as the essential variable in the study of change in other life experiences is used in two clearly distinguishable ways, though it is easy to confound them because they both use the concept of stages. In the dominant approach stages are age-dependent. In the second usage stages are distinguished in the experience of relatively enduring life situations, such as marriage or careers or unemployment without regard to the age of the people under study.

Once again, it is Piaget and Freud who have made the most influential contributions to the identification of age-dependent stages in development, even though both are now the subject of much controversy.

Piaget suggests several consecutive stages in cognitive development from infancy to adolescence which have to be acquired in an invariable order. His concept of stages has, however, now come into question. It is said to imply too orderly a sequence, leading to the assumption of discontinuous mental states. The critique of Piaget in this respect is substantiated by experiments demonstrating that the human infant is already at birth equipped with a more competent mental structure than he assumed and by reference to the all-too-obvious fact that adult reasoning is in many situations less logical than the highest stage would lead one to expect.

I am not sure that Piaget actually thought of his stages as discontinuous, even though some of his formulations suggest it. His insistence on the continuing oscillation between assimilation and accommodation implies gradual change, much in line with Jaques's formulation that time must be thought of as the continuity of existence, that change can only be recognised if the original and the subsequent state have something in common.

The debate about the Freudian developmental stages is utterly confused by a widespread semantic misunderstanding of what he meant by sexuality which I have tried to clarify elsewhere (Jahoda, 1977). Here it must suffice to note the special features of the Freudian view of development: there are here no discrete stages, but a gradual unfolding. The early pleasure zones are not replaced but added to as children progressively explore their own bodies in interaction with their caretakers.

Notwithstanding the controversies about the content of stages and about their formal aspect – discrete substitutions or continuous unfolding – the idea of stages is central to the work of many time-conscious students of life events. So

Lawrence Kohlberg (1963) has, over the last quarter of a century, applied for an extended age range the Piagetian notion of stages in the development of moral judgment. He has convincingly demonstrated that the rules underlying judgments in moral dilemmas (e.g. what is worse, accidentally dropping three plates or deliberately smashing one? or: is stealing to buy medicine for one's sick spouse right or wrong?) change with age. While this is uncontested there is, once again, much debate about the invariable sequence of the postulated stages. Kohlberg seems to have imposed the value of individualism, so deeply rooted in American culture, on his hierarchy of moral judgments. His highest stage consists of everyone judging autonomously according to personally established principles, yet all agreeing on principles which meet the interest of the collectivity to which they belong. Drew Westen (1985) has recently discussed Piaget's and Kohlberg's work on moral development and suggested that autonomy and consensus are incompatible.

Perhaps the most telling critique of Kohlberg's proposed invariable sequence of stages comes from Carol Gilligan (1982). She does not question that the application of abstract principles may be the highest stage men reach in moral judgment, but she does show that the other half of mankind functions differently. Women's mature moral judgment is closer to the reality of human relations than to principles. Will it be so in the future or will men's moral hierarchies change? We do not know.

Beyond infancy and childhood some attempts have been made to discover psychological regularities for the entire span of human life. Based on psychoanalytic ideas Erik Erikson (1950) has extended and modified the Freudian stage concept. His age range goes up to adulthood in eight steps. The psychological meanings of each stage are presented as alternative, e.g. trust versus mistrust for infancy, ego integrity versus despair for maturity; an individual's adoption of one or the other psychological stance depends on inner regularities as much as on influences and events from the outside. Unless the challenge presented at one stage is successfully met, progress to the next is inhibited; the acquisitions during earlier stages remain the foundation for later ones.

A different view on life-span development was adopted by

Charlotte Bühler (1933) who explicitly related psychological and biological stages in her biographical studies of human lives. She divided the biological curve into five periods: attainment of sexual maturity by about the age of 15; attainment of full height and weight by about 25 years, followed by about 20 years of full biological maturity; then 10 years of gradual decline in sexual functioning which leads at about 55 to the end of sexuality and the onset of old age. These somewhat arbitrary date lines (proposed by a then relatively young woman) cannot be taken literally. However, wherever one puts the markers, some such biological curve with psychological parallels does occur. Bühler studied the lives of the great during the preceding 200 years for whom biographies were available (Goethe, Wagner, etc.) along three psychological dimensions: experience, behaviour and achievement. She discovered certain similarities in experience and to some extent in behaviour across lives at different historical periods in line with the biological curve. Since Goethe was the prototypical example, it is hardly surprising that the experience in the second stage was described as *Sturm und Drang*, the behaviour as trial and error. There were however, no regularities with regard to achievement peaks which came earlier in some cases, later in others than Bühler had expected.

Quite recently interest in the study of the entire life span has reawakened in the United States, where it has been predicted that during the next decade such studies will emerge as a new multidisciplinary field in the human sciences.

Controversy is also in evidence with regard to the second usage of the stage concept where it is applied to the study of the psychological consequences of prolonged periods of unemployment. Broadly summarising research from the 1930s Eisenberg and Lazarsfeld (1938) conclude that there is a systematic sequence where the initial shock experience is followed by some recovery and active job hunting which, if unsuccessful, leads to distress and in the end to fatalism and apathy. The descriptive adequacy of this sequence has been criticised even for the 1930s (Fryer, 1985) and modern research has failed to confirm it.

One difference between early and contemporary studies is the length of the unemployment period investigated; two years and more then, at best one year now. This second usage of stage is, however, generally restricted to the enduring Now without regard to measured time-spans or age. There are, of course, many modern studies that specify the age of the unemployed in their samples but as a rule they leave it to the reader to infer the degree to which the prolonged experience of unemployment is influenced by the interaction of enforced idleness and age specific needs at various life stages. Whether or not the early suggested sequence of adaptational stages holds today for certain age groups but not for others remains to be established. That biologically linked age-specific life-styles and expectations influence adaptation to external events is beyond question. It has been demonstrated in the field of unemployment research explicitly, albeit at the expense of ignoring the adaptational sequence. Using Bühler's life-stage concept as a guide it could be shown that the psychological benefits derived from a scheme that provided regular work and some financial support for unemployed miners in South Wales differed systematically for different age groups (Jahoda, 1938/1987). Everybody derived some economic advantage from membership; but the psychological benefits occurred by and large only to those in life-stages four and five.

Systematic stage sequences within an enduring Now of a prolonged life situation have also been established for other experiences. When the situation extends over very long time periods, as in the study of professional careers, age is at least tacitly implied. The now much debated stage of 'burn-out' is not a phenomenon of life stage two or three (following Bühler's stage division). Jaques's identification of a 'mid-life crisis' (Jaques, 1965) is explicitly placed around the end of a person's fourth decade.

Other stage sequences, for example in the study of friendships (Duck and Gilmour, 1981), are presented without reference to the age of the friends; the empirical data which support the suggested sequence have as a rule been obtained from young people only.

As the examples show the two concepts of stage are indeed distinct in the manner they deal with time. It is the second one that could achieve an even closer approximation to lived reality by greater sensitivity to the biological life curve. The study of the experience of time as much as the use of time as an essential variable in the identification of developmental stages assume implicitly – and sometimes explicitly – that the regularities so discovered have transhistorical validity. This assumption has recently come under sharp attack.

So Janet Spence (1985) in her presidential address to the American Psychological Association has this to say:

> . . . many psychologists . . . appear to assume that their research findings have . . . universality. . . . This interpretation is probably valid in the case of many relatively basic phenomena. However, an ahistorical, acultural approach also characterises many investigations of complex social phenomena . . . But it seems indisputable that many of our empirical findings . . . must inevitably be bound to particular cultures and times . . . (Spence, 1985, p. 1285)

Kenneth Gergen, a social psychologist, goes even further (Gergen, 1973 and Gergen and Gergen, 1984). He argues that at least this branch of the subject should be explicitly recognised as contemporary history; he blames logical empiricism for having numbed the sensitivity of psychologists to historical change which is non-recurring and subject to luck and chance. This wholesale disclaimer of transhistorical validity has been received with unease by many social psychologists; others have rejected it outright, but the majority continue to ignore historical change.

The reason for this mixed reception lies in the relation between mainstream psychology and its unruly offspring, social psychology. Psychology qua natural science has produced a truly staggering amount of knowledge about psychophysical processes, particularly about perception, memory and learning, notwithstanding the diversity of theories in these areas which differ essentially in their assumptions about whether these processes are innate or learned. Whatever the explanation, there is agreement about some basic perceptual phenomena, for which the assumption of independence from historical change is exceedingly plausible.

Matters of memory and learning are more complex because meaningful content is as a rule involved and meanings are subject to change over time. But even in perception there is at least one process which in western cultures has all the earmarks of timeless universality in that it is involuntary and automatic, but emerges as culture-bound, by implication the result of differing histories. I refer to perceptual illusions, particularly the Necker cube and the Ponzo illusion. Richard Gregory (1972) writes that the Zulus see the Necker cube as a flat figure and do not succumb to the size distortion in perspective drawings because their natural environment contains no straight lines: they live in circular huts. Such illusions are thus obviously subject to change in the course of historical processes. This exception is worth noting because it indicates that not even in a natural science approach can timeless universality for some phenomena be taken for granted.

At the beginning of this century the superiority of the natural science approach in psychology was virtually beyond question. Even Freud thought of himself as a natural scientist engaged in the discovery of timeless phenomena. Though he had so much to say about time in all its dimensions, he was blind to the possibility that the psychological manifestations he observed were themselves the result of a limited historical period. But the symptoms of the hysterical Viennese middle-class women whom he treated have by now virtually disappeared; some of his statements about women in general are more in line with the then dominant stereotypes than with psychoanalytic ways of reasoning. Whether other phenomena for which he claimed universal validity can be so regarded is an open question. For the span of more than half a century two lines of evidence are available. First, nearly 1000 experiments in western cultures have been performed to see whether phenomena such as repression, wish-fulfilling dreams, castration fears, oedipal involvement, etc. could be demonstrated; second, cross-cultural research on these matters has been conducted. Both methods have been used and misused to confirm and disconfirm (these two dichotomies to be understood as operating in parallel *and* cross-cutting). But the controversy is far from over.

Academic psychology, less concerned with content and

meaning and more with formal regularities, has largely adhered to the natural science paradigm, produced its achievements and acquired its prestige by doing so and has passed on to social psychology this scientific credo and the corresponding methodology.

This heritage sits uneasily with the central task of social psychology which it shares with other social sciences: to elucidate the relation between individual and society in their ongoing interaction where changes in society have repercussions in individuals. There is here a dilemma: only one partner in the relationship, people, can be subjected to natural-science-type experiments with their search for timeless universality, society cannot; both, however, need to be understood.

Many social psychologists have dealt with the dilemma by adopting so narrow a conception of the social that it could be regarded as ahistorical. If the concrete content that every experimenter must construct in the study of perception or judgment is a social, not a physical object, as in person perceptions, e.g. it is regarded as social psychology. The same is true if more than one person is simultaneously exposed to an experimental manipulation. Small groups, often constituted for experimental purposes only, belong to this category where results are as a rule reported as averages of individual reponses. Since it has been demonstrated that people respond differently when isolated and when in the physical presence of others there is here a social component. It is also legitimate to make the assumption that even isolated individuals carry within them the stamp of the society to which they belong. Much excellent psychological research has been done within such narrow and unanalysed definitions of the social. But, of course, in that sense all psychology is social psychology. The experimental situation itself is a social situation, albeit a peculiar one. Trouble, however, often arises when the psychological phenomena so identified are transferred to the world outside the laboratory. More often than not such transfers prove to be impossible because unanalysed social factors produce differing results. The social is more than implied in the recognition that our Selves are socially constituted.

But even within a narrow conception of the social, difficulties with historical time have been demonstrated through the all-too-rare replication of experiments after one or more decades.

There are some examples, however. In 1966 Waly and Cook replicated a study on the impact of attitudes on learning and forgetting; since the original experiment in 1943 had clearly demonstrated that people recall best matters congruent with their own attitudes that had been accepted as a relatively enduring regularity, was confirmed in 1958, but by 1966 the phenomenon seemed to have disappeared.

Similarly, the well-known conformity experiments in the early 1950s by Solomon Asch (1951) have apparently not stood the test of time in the 1980s (Perrin and Spencer, 1981). Asch had presented experimental evidence that about one third of the people in various groups yielded in their judgment about the lengths of lines to the judgment of their peers who had been instructed to give the obviously wrong answer. As with all experiments his findings were presented within a timeless here-and-now perspective. The time-boundedness of the phenomenon has now been established.

While experiments are normally replicated in the expectation of confirming earlier results, the repetition of survey questions over the years is intended to demonstrate change. Survey research, customarily regarded as the intellectual property of sociology, seems to me to be de facto sharing this narrow view of the social in experimental social psychology. The interesting phenomena it reveals are people's thoughts, opinions, intentions and feelings. The social world implicated in these changes remains, however, unanalysed. Traditional categories such as social class or level of education are used in the analysis of individual responses but external changes which are partly responsible for internal ones remain unanalysed.

Nonetheless, the replication of surveys over an interval of years produces interesting results, clearly demonstrating that changes must have occurred. The surveys leave it to the imagination of the reader to construct their own social history of a past decade or two to make sense of the reported changes. So Joseph Veroff (1984) from the Survey Research

Center in the States, e.g. compares the results of surveys conducted on identical issues in 1957 and 1976. To take an example, confirmed by others, trust in government was less frequent at the later stage. Whether people had become more sophisticated and distrustful or whether government practices had changed, or both are questions unanswerable from survey data which deal with individuals, not with institutions and their change. Like experiments, surveys are by their very methods compelled to operate in the here and now. Only replications reveal that yesterday's Now may have changed and requires attention both on the individual and the social level.

Thus, the results of replicated experiments and survey questions confirm that the natural-science stance in social psychology, notwithstanding its prolific output, fails to establish what it aimed for: the discovery of regularities impervious to historical change. However justified the assumption of timelessness is in mechanisms of perception or in the dance of particles, it is illegitimate in studying the interaction between people and society. If this is the aim, the assumption that psychological phenomena are determined by the enduring equipment of the human organism alone is as untenable as its opposite that they are exclusively the product of changing social events. Of course it is human actions that change the social world; but we are born into one made by our ancestors over many generations. While every generation produces some changes, none can undo the past as it continues in the social present with its unknown duration.

If this is no more than common sense, why is it so difficult to do justice to it in social psychological research? There are several reasons, among them the differing intellectual traditions in the two constituent disciplines, psychology and sociology, from which social psychology derived. But in the present context something else is important: social change has a speed, rhythm and rate of its own, uncoordinated to individual change experiences and yet influencing them deeply. It is easier in research to keep at least one of the two constituent parts static and so settle for snapshots of the interaction whose transhistorical validity is thus left open to question.

While there is much talk about future shocks and the increasing rate of change in complex societies, how to define rate of change is an as yet unsolved problem in the social sciences for good reasons. The social structure consists of an untold number of formal and informal institutions: economic, political, educational, health, legal, art, science and religious institutions as formally constituted, with their regionally distinct traditions and specific propensities to change in one or more of their features. Family, marriage, social class, norms and customs are examples of informal institutions. Both types of institution are subject to change in historical time-spans. Some undergo gradual, hardly perceptual change, others are deliberately and abruptly altered so that change is discontinuous. Which of these should enter into an overall rate of change is indeed difficult to decide, perhaps undecidable in a generally acceptable way. By and large social scientists have preferred to study single institutions rather than this bewildering multiplicity, but even within these limits rates of change remain largely undefined.

Those who aim for results of more than contemporary validity are perhaps better advised to concentrate not on change but on the relatively enduring aspects of people and of institutions. For reasons indicated before the search for what is relatively enduring needs to proceed on two separate tracks, comparing phenomena over an extended stretch of time so that the fit or misfit between psychological needs and social arrangements can be assessed. This is easier said than done. However, a tragic opportunity to do just this arose over a comparison of the studies of mass unemployment in the 1930s and now. The social world has changed in this half-century almost beyond recognition, but the psychological response to unemployment shows many similarities, particularly in the experience of empty time. Surely something must have remained relatively constant to produce similar results under such different general conditions. Retrospectively it becomes clear that employment as an institution whose time structure shapes the entire way of life of an industrial society has not changed; neither has the need of people for institutionally supported time experiences. One may resent the power of employment as an institution to impose its time structure on time experience; even suspect that its dominance has destroyed in many the ability to structure their own time use. But this is what social psychology can only reveal; it cannot change it.

To sum up: the treatment of time in social psychology is complicated by the inherent difficulty of its major task which consists of analysing both people and social institutions, each with their more or less enduring features. Much of social psychology can therefore claim only contemporary validity. But it is possible to distinguish the relatively more from the relatively less enduring and so produce social psychological insights that can claim more than momentary validity, if not for always so at least for yesterday and tomorrow.

Finally the question arises of what the relative neglect of historical time in psychology and social psychology says about the intellectual standing of these disciplines. Given that the experience of time is universal and of overwhelming significance, can studies that neglect this issue be taken seriously at all? I believe they can and must. Psychology *qua* natural science is legitimately engaged in the search for human universals. Most of social psychology does not go beyond establishing psychological phenomena in the here and now and thus is only a contribution to contemporary history. But this in itself is worthwhile for the understanding of the world in which we live. Such studies are the bread and butter of the psychological research enterprise; not they but their sometimes claimed and sometimes tacitly assumed timeless validity must be rejected.

Many psychologists prefer to concentrate on the study of fundamental formal processes which, when they are not dependent on content and meaning, can be assumed with some justification to be relatively enduring. On this basis, formal questions can be raised and answered without concern for time dependency. From this point of view human nature does not change; from the equally legitimate one of social psychology it varies with historical circumstances. The closer a research question is to psychological and social reality, the less justified are assumptions of timelessness. Social psychology has, therefore, a more difficult task in this respect than its parent discipline because speed and rhythm of change in the social world are uneven so that the old and the new coexist (Victorian values at the end of the twentieth century), are not easily and certainly not uniformly perceived, and uncoordinated to personal time experience at different stages of the life cycle. Perhaps it is unrealistic to expect that contemporary social psychology could proceed by taking note of the complexity of time in all its dimension. The discipline is too young to provide more than selective snapshots through the present. In a hundred years or so our intellectual heirs may be in a better position to identify relatively enduring social psychological phenomena. Only time will tell.

REFERENCES

Asch, S.E. (1951), 'Effects of group pressure upon the modification and distortion of judgements', in H.S. Guetzkow (ed.), *Groups, Leaderships and Man*, Pittsburgh, Carnegie Press.

Bühler, C. (1933), *Der Menschliche Lebenslauf als Psychologisches Problem*, Leipzig, Hirzel.

Duck, S. and Gilmour, R. (1981), *Personal Relationships*, London, Academic Press.

Eisenberg, P. and Lazarsfeld, P.F. (1938), 'The psychological effects of unemployment', *Psychological Bulletin*, vol.35, pp. 358–90.

Erikson E.H. (1950), *Childhood and Society*, New York, W.W. Norton.

Fryer, D.M. (1985), 'Stages in the psychological response to unemployment: a (dis)integrative review', *Current Psychological Research and Reviews*, vol. 4, no. 3, pp. 257–73.

Fryer, D.M. and Payne, R.L. (1984), 'Proactivity in unemployment: findings and implications', *Leisure Studies*, vol. 3, pp. 273–95.

Gergen, K.J. (1973), 'Social psychology as history', *Journal of Personality and Social Psychology*, vol. 26, pp. 309–20.

Gergen, K.J. and Gergen M.M. (eds) (1984), *Historical Social Psychology*, Hillsdale, N.J., and London, Erlbaum.

Gilligan, C. (1982), *In a Different Voice*, Cambridge, Harvard University Press.

Gregory, R.L. (1972), 'Visual illusions', in B.M. Foss (ed.), *New Horizons in Psychology*, I, Harmondsworth, Penguin,

pp. 68–96.
Gruber, H.E. and Vorrèche, J.J. (eds) (1977), *The Essential Piaget*, New York, Basic Books.
Jahoda, M. (1938/1987), 'Unemployed men at work', in D.M. Fryer and P. Ullah (eds), *Unemployed People*, Milton Keynes, Open University Press.
Jahoda, M. (1977), *Freud and the Dilemmas of Psychology*, London, Hogarth.
Jahoda, M., Lazarsfeld, P.F. and Zeisel, H. (1933), *Die Arbeitslosen von Marienthal*, Leipzig, Hirzel.
James, W. (1890), *Principles of Psychology*, vol. 1, New York, Henry Holt & Co.
Jaques, E. (1951), *The Changing Culture of a Factory*, London, Tavistock Publications.
Jaques, E. (1956), *Measurement of Responsibility*, London, Tavistock Publications.
Jaques, E. (1965), 'Death and the mid-life crisis', *International Journal of Psychoanalysis*, vol. 46, no. 4, pp. 502–14.
Jaques, E. (1982), *The Form of Time*, London, Heineman, New York, Crane Russak.
Kohlberg, L. (1963), *The Development of Children's Orientation Toward a Moral Order*, Vita Humana, Basel.
Novotny, H. (1975), 'Time structuring and time-measurement: on the interrelation between timekeepers and social time', in J.T. Fraser, and N. Lawrence (eds), *The Study of Time*, *II*, Berlin, Heidelberg, New York, Springer Verlag.
Perrin, S. and Spencer, C. (1981), 'Independence or conformity in the Asch experiment as a reflection of cultural and situational factors', *Britsh Journal of Social Psychology*, vol. 20, pp. 205–9.
Spence, J.T. (1985), 'Achievement American style – the rewards and costs of individualism', *American Psychologist*, vol. 40, no. 12, pp. 1285–95.
Veroff, J. (1984), 'A changing character of cultural dispositions: a social indicators approach', in K.J. Gergen and M.M. Gergen (eds) (1984), *Historical Social Psychology*, Hillsdale, N.J., and London, Erlbaum.
Waly, P. and Cook S.W. (1966), 'Attitude as a determinant of learning and memory: a failure to confirm', *Journal of Personality and Social Psychology*, vol. 4, pp. 280–8.
Westen, D. (1985), *Self and Society*, Cambridge, Cambridge University Press.

CHAPTER 8

TIME, HABIT AND THE FRATERNAL IMPULSE

Johnston Birchall

> It is evident that life in societies would be utterly impossible without a . . . development of social feelings, and, especially of a certain collective sense of justice, growing to become a habit. (Kropotkin, 1915, p. 58)

If the building up of sociological theory had proceeded in as orderly a manner as Durkheim and Weber hoped all the concepts thrown up by them would by now have been examined exhaustively and fitted into an overarching theory. Yet we know that this has not happened, that instead of a patient building on classical foundations there has been a partial accumulation of knowledge within limited world views, which has been achieved at the expense of the over-stressing of some concepts and the overlooking of others (see Kuhn, 1962). Like the unfortunate health authority officials who were reported recently as having built a multi-million pound general hospital while forgetting to include a kitchen, sociologists seem to have overlooked some concepts vital to the application of their theories. In this paper I want to draw attention to three such concepts: social time, which, since Sorokin made a major contribution to the field in the 1940s, has been strangely neglected; habit, which formed a key part of both Durkheim and Weber's theories but has since been ignored by sociologists if not by psychologists, and thc political philosophers' concept of fraternity, which in contrast to the other two parts of another famous trilogy – liberty and equality – has almost dropped out of sight. Yet the above quotation from Kropotkin suggests that these concepts might not only be central ones in their own right, but might also suggest solutions to a major sociological problem concerning the human capacity for mutual co-operation. Kropotkin links the 'development of social feelings' (social solidarity), more specifically 'a certain collective sense of justice' (fraternity), with the notion of habit and the central social scientific problem of 'life in societies' (social order).

The purpose of this paper is to explore these connections through quarrying the work of some major theorists, in the hope that further, more painstaking work will follow. Firstly, each of the concepts is explored and suggestions are made about why they may have been neglected. Secondly, connections are made between the concepts of habit and fraternity which suggest that each one can only be understood fully if the other is taken into account. Thirdly, in order to find out how habits begin, are sustained, and change over time, and under what circumstances they generate fraternity, a typology of social exchanges is introduced which ranges from pure theft through more or less equal exchange of equivalents to the pure gift. Fourthly, by means of this typology a 'co-operative mutualist' theory of exchange is outlined which brings out the full force of Kropotkin's linkage of habit with fraternity, and it is contrasted with a deliberately negative theory (a kind of null hypothesis) labelled 'competitive individualist'. Fifthly, Axelrod's recent work on games theory is examined. This provides evidence which both strengthens the mutualist viewpoint and specifies more exactly the conditions under which habitual exchanges generate fraternity. Finally, this allows a critical assessment of Durkheim's proposition that the exchanges necessitated by the division of labour automatically generate fraternity; the fraternal impulse can be expected to 'grow to be a habit', but only under certain specifiable conditions.

THE NEGLECTED CONCEPTS: TIME, HABIT AND FRATERNITY

In a forthcoming book Michael Young argues that sociology

should take seriously the concept of time, and that 'time sociology' should be developed as a sub-discipline comparable to time geography and chronobiology (Young, 1988). There are two ways of looking at time: the linear and the cyclical. Linear time is familiar, conceived as the line extending into the future along which social change is measured, but cyclical time, conceived as a set of recurrences which extend back into the past and preserve social continuity, has been comparatively overlooked. The reason is that the balance between the linear and cyclical world views varies from one society to another. Sociologists are often as submerged in their taken-for-granted everyday world as are lay members, and in this 'metronomic society' the emphasis on linearity and social change is certainly taken for granted; in fact the measurement of mathematical time has reached such a peak of accuracy and ubiquity that it seems the only way of conceiving of time. However, the cyclical is at least as fundamental, being based partly on natural astronomical cycles such as the day and year, and partly on socially constructed ones such as the week. Furthermore, as well as being time-locked, they can be sequence-locked, that is, based on routine sequences which are entrained whenever a similar set of circumstances is encountered, regardless of the time of day or year. The cyclical is that which we do again and again in more or less the same way or at more or less the same time.

Such cycles can also be called by the rather old-fashioned name of 'habits'. The meaning of habit has remained fairly constant over time. It refers to actions that are 'relatively unmotivated' (Giddens, 1979, p. 218), or more positively, to 'a more or less self-actuating disposition or tendency to engage in a previously adopted or acquired form of action' (Camic, 1986, p. 1044). Camic suggests a useful typology ranging from simple to complex habits. Simple ones are virtually automatic skills such as writing or driving a car. In the middle range they include habits of interpersonal interaction, economic, political or religious behaviour, habits of obedience and so on; at this level concrete cases can also be more or less habitual or reflective, but to be called habits the habitual part of them has to predominate. Finally, complex habits approximate to a person's whole disposition or character. Young connects it more explicitly with time, describing a habit as 'an expression of one of the most fundamental cyclical devices of the mind, employed by it on a large scale to take experience out of the flow of time' (1988, p. 94).

One might ask why, if it is so fundamental, the concept has been neglected by sociologists for so long. Young suggests that habits are subject to 'the law of the disappearing cycle' whereby they are literally forgotten, become unavailable to scrutiny by the reasoning mind, and so are a difficult subject for study. Camic points out also (1986) that habit became a 'lapsed alternative' within sociology not because of any deficiency in the concept (since it was a major background factor in the writings of both Weber and Durkheim), but because of an unfortunate split between intellectual fields which led sociologists to over-emphasise reflective action in a revolt against behaviourist psychology.

Fraternity is more difficult to define, because it is as Durkheim puts it (referring to its synonym social solidarity) 'an intangible phenomenon which does not lend itself to observation' (1964, p. 67). Like habit, it shows up better in its effects, appearing positively in the form of laws (we might from Young's work add customs or collective habits) and negatively in social indicators such as suicide rates. Just as habits are buried in the unconscious, fraternity is buried in the intuitive feelings of the individual to others, feelings of what Kropotkin calls 'compassion'. Those few political theorists who have used it tend to play it down; Perry calls it equality of esteem or mutual respect, and insists that it does not imply friendship or real intimacy, but courtesy, general friendliness, which though it underlies familial, neighbourhood and work ties can be generalised to citizens as a whole, even to all of humanity (Perry, 1944). Pennock goes further, calling it 'the spirit of brotherhood, of treating others not simply as though they had rights equal to ours, but with living concern for their welfare' (1950, p. 94). If it is not just a question of rights though, it is not really a political concept at all, but a kind of excess of feeling which arises out of, but does not add to, arguments for equal democratic rights. He is therefore able to dismiss it quite quickly while leaving behind at the last moment a potential argument which the sociologist, if not the political theorist, might be interested in picking up. He calls fraternity a spirit without which the value of democratic rights would soon be lost sight of, and suggests that its function is to promote mutual adjustment among individuals when their rights collide. In other words it maintains and arbitrates between liberty and equality. Or as a sociologist might put it (calling it social solidarity), it maintains common values or norms and, when these are in conflict, provides a general willingness to compromise.

A.H. Halsey clearly makes this connection when in describing British society he talks of 'an unresolved problem of fraternity or basis for social order, beneath the clash of egalitarian and libertarian argument' (1978, p. 10). He complains that though the concept of fraternity has been left to moralists and theologians to deal with, it is sociologically prior to liberty and equality. It is 'the principle which defines the group within which certain social values are to be applied' (Halsey 1978, p. 11). Liberty and equality only arise as demands because there is a group which, because of the fraternal impulse, already exists. Their competing claims can only be resolved without social conflict and the possible destruction of the group if that same impulse is allowed to arbitrate. This still sounds a little vague but John Rawls makes it clearer. He argues that as it stands fraternity expresses no definite requirement: 'We have yet to find a principle of justice that matches the underlying idea' (1973, p. 105). His 'difference principle' does correspond to the natural meaning of fraternity though, as 'the idea of not wanting to have greater advantages unless this is to the benefit of others who are less well off'. Egoism is thus checked by a deeper sense of mutuality. Against the argument that ties of sentiment and feeling are unrealistic in a large and complex society, he argues that fraternity, when transposed into this difference principle, can be made a perfectly feasible standard which imposes a definite requirement on the basic structure of society.

Again, if, like habit, it is so important, we might ask why the concept has almost fallen out of use. Theorists have repeatedly (one might say habitually) built up, defended and modified complex political philosophies in terms of the competing claims of liberty and equality, while as R.H. Tawney says 'Fraternity . . . has hardly been considered at all' (1964, p. 164). When it has been considered it has been reduced to the much tamer concept of citizenship (Marshall, 1977). It may be that, just as habit was associated with a discredited psychological theory – behaviourism – fraternity was associated for a time with a discredited (if that is a strong enough word) political theory – fascism. Perhaps also, just as habit became the exclusive territory of the psychologists, fraternity, in the form of social solidarity, was captured from the political theorists by sociologists; these are questions which can only be answered by a historical exploration similar to that which Camic has completed for the concept of habit. Or it may be that in the wider society, just as the prevailing emphasis on linearity has driven out the cyclical, the emphasis of the two main competing political theories of state socialism and free market capitalism have emphasised equality and liberty while driving out that much less definable value – fraternity. Or more likely even than these explanations may be a simpler one; perhaps both concepts are so fundamental that we have (habitually) grown to take them for granted.

THE CONNECTION BETWEEN HABIT AND FRATERNITY

Is there a connection between these two concepts other than the trivial one that they both appear to have been neglected? The clue seems to be that they are both intimately connected to the vital concept of social order, of how a society holds together in the first place. Firstly, Young claims that the cyclical is 'one of the most elemental of the forces in the social structure' (1988, p. 19). For a society to exist at all, the behaviour of its members has to be some extent mutually predictable, yet if everything is changing quickly in a linear fashion such predictability must be problematic. Habit tends to slow down this linearity, preserving the past by bringing it continually into the present, and enabling people to synchronise their activities according to repeated common cycles.

It thrives on similarity rather than difference and so is a conservative force for continuity against change. Furthermore, it evokes sentiments of social solidarity or fraternity among those who practise common habits or customs. Similarly, Camic points to Durkheim's view that habit is 'one of the principal supports for the moral fabric of modern societies', and Weber's view that habits (or customs) are frequently turned into binding norms by the mere fact of their regular recurrence (1986, p. 1039, p. 1059).

If habits held in common act to create fraternity, then the opposite is equally true that fraternity creates habits. Kropotkin talks of a sense of justice 'growing to be a habit', and Durkheim echoes this when he says that moral behaviour is easily transformed 'into habit mechanically carried out' (Camic, 1986, p. 1051). On the other hand Durkheim is worried that, once they are fixed, habits can be a great impediment to change; there can be both good and bad habits, and the moral reformer has to make sure that the good ones are implanted before the bad can take a hold. For this reason he stresses the fixing of moral habits by primary education and co-operative occupational groups. Similarly, Weber acknowledges that much of social action is governed by habit rather than rational reflection. Traditional action, as action 'determined by ingrained habit', forms 'the great bulk of all everyday action', and as such it both reflects and creates traditional authority and custom (Camic, 1986, p. 1060).

Granted then that there may be a close relationship between the disposition to habitual behaviour and the feeling of fraternity, is it possible to specify more closely how they interact and whether there are other intervening variables which might affect their relationship? It is just possible to imagine a society in which people act according to deeply ingrained habits which they duly pass on to their children, and which enable them to synchronise and co-ordinate tasks, but which leave them with no particularly warm feelings towards each other. Cohen makes some crucial analytical distinctions between social solidarity (which we might for the moment treat as a sociological synonym for fraternity), consensus, commitment, cohesion and so on (1968, ch. 6). These suggest that a society may be cohesive without being particularly solidary, and vice versa. There may be functional integration between people through a highly developed division of labour, but this need not (as Durkheim assumed) lead to feelings of solidarity and the development of group norms.

For this reason Kropotkin points to an intervening variable, the innate sociability of the animal (1915); it is impossible for a social animal such as *homo sapiens* to interact with others and not to develop feelings of solidarity. Simmel confirms this view sociologically by observing the generation of such feelings in that most basic of all social forms, the dyad, the face-to-face interaction of two people, which he sees as the 'germ of sociation' because it has the quality of 'pure and immediate reciprocity' (1964, p. 136). This reciprocity produces what he calls 'gratitude', but what could equally well be termed fraternity or social solidarity. Whatever we call it, it 'establishes the bond of interaction'. It is interesting that he calls gratitude the 'moral memory of mankind', because (in Young's theory) memory is the other great recycler of linear time. It would be more correct though to call it the moral habit of mankind, because Simmel goes on to explain that it is really more of a 'creative mood' than a memory, a sense of solidarity that is triggered off in situations which are similar to previous situations in which a reciprocal giving had produced mutally beneficial effects. Gratitude may only be a feeling or mood, but its effects are far from intangible; in Simmel's view 'its thousand-fold ramifications throughout society make it one of the most powerful means of social cohesion' (1964, p. 388). Reciprocity creates fraternity, then, and habit, by reawakening the feeling of fraternity, leads to the repetition of those kinds of reciprocal relationships which have proved fruitful in the past; habit creates fraternity, and fraternity creates habits. In practice the two are indissolubly fused. In asking ourselves whether or not to enter into an exchange, we need not recall similar exchanges which have taken place many times before, but need only take note of the feeling we get from being in that situation. If we glow with gratitude the exchange can safely be made, if we shake with rage it had better be left alone. On the other hand, when a habit becomes deeply entrenched, the exchange may occur before any particular feelings are invoked, the habit leading and the feeling following along behind.

Such an argument is in sociological terms quite robust. For instance, an arch-rationalist might argue that this fusion of habit and feeling is in a market society unnecessary, that reliance on price-signals alone is necessary for a rational calculation of one's utility. George Homans has expressed this rationalistic attitude to exchanges in the following proposition (1974, p. 45). 'In choosing between alternative actions, a person will choose that one for which, as perceived by him at the time, the value V of the result, multiplied by the probability P of getting the result, is the greater.' However, the probability of getting a result depends partly on previous experiences, and so the glow of gratitude which Simmel notes may be as rational a response as any. It may be a kind of sub-rational store for the price-signals of the past, a way of weighing up the quality of the intended interaction by testing it against the extent to which it matches previous experience. Alternatively, an arch-behaviourist might also argue that, because of the primacy of the subterranean promptings of habit, feelings of fraternity are unnecessary. Some of the behaviourist propositions support this view. For example, the 'success proposition' states that the more often an action is rewarded the more likely a person is to perform it, and the 'stimulus proposition' states that if stimuli associated with previously rewarding situations are present, they will induce a person to perform a similar action. However, Homans states in behaviourist terms exactly the same point as Simmel, that if an action receives the expected reward, or a greater one, a person is pleased, and is more likely to perform approving behaviour. Even when reduced to a habituated individual intent only on obtaining rewards, theory has to take account of the way in which social exchanges seem to generate fraternity.

Is there a macro-sociological way of looking at the connection between fraternity and habit which transcends the essentially individualistic exchanges of exchange theory? R. H. Tawney equates fraternity with what he calls 'common culture', a sense of common interest among people which transcends the individual's immediate self-interest and forms a different, though no less rational, basis for action (1964,

p. 108). This approach begins not from the individual in the one-to-one situation of simple reciprocity, but from society as it is already constituted to the individual. It suggests that people are socialised into a society which already contains common ways of looking at the 'every-day world' (Berger and Luckmann, 1971), ways which are inherited by the individual as a comprehensive world view so fundamental that it is taken to be reality itself. Berger and Luckmann talk of 'habituation', and this dynamic term for habit emphasises that it is not just something which builds up anew from each individual's experience, but is transmitted culturally as a given. Fraternity, on this macro-sociological level, exists to the extent that the everyday world transmitted to the individual emphasises common needs, goals and responsibilities. People will treat each other as being equal in certain fundamental respects not because of rational calculation or even habits born out of personal experience, but out of a sense of common culture, Kropotkin's 'collective sense of justice, growing to become a habit' (1915, p. 58). Not to do the habitual action brings automatic condemnation from one's fellow citizens, and to do the opposite leads to the questioning of whether the actor is a citizen at all, whether he has put himself 'beyond the pale'.

However, though this way of looking at the subject may be sociologically more realistic, setting the individual within a given society, it is not as effective as the more individualistic view at showing how habits begin for the first time or undergo change. It may be useful then to pursue further the connections between these two partly forgotten and obscured concepts – fraternity and habit – through what Simmel calls 'reciprocity': the medium of exchange theory.

A TYPOLOGY OF EXCHANGES

It is impossible at this point to enter into a detailed critique of exchange theories, and so only some minimal observations will be attempted, and then two types of theory will be suggested which encapsulate, in an economic but by no means exhaustive way, opposite approaches to the subject. Firstly, in trying to understand social and economic exchanges one has really to separate out two variables: the forms which exchanges take, and the moral quality of the interaction between the people involved. The relationship between these two variables is complex and can only be understood when we bring in the crucial intervening variables: linear and cyclical time. The forms of exchange can be defined along a continuum, the polar opposites of which are pure taking and giving, theft and the gift. Of course, these are not technically exchanges at all, but one-way transfers which signal the end points at which exchange turns into something else. In between, in the exact centre of the continuum, there is that equal reciprocal exchange which in classical economic theory is the ideal form achieved in a free market. The virtue of such an image is that it turns the forms of exchange from sets of more traditional concepts, which tend to confuse the forms with their moral content (altruism, egoism and so on), into a continuous variable which allows exchanges to be placed more exactly, according to what is actually being given and received.

For instance, an altruistic gift can be placed at the 'pure gift' end of the continuum only if it is entirely disinterested, in the Buddhist sense of an act of pure compassion. More likely there is some form of exchange, if only of gratitude on the part of the receiver. If the giver values this gratitude, and gives in order to receive it, then the exchange may slip towards the centre of the continuum and become a more equal exchange. In societies where charitable giving is sanctioned by religious belief, the exchange may become an equal one, with the giver giving in order to receive in return an eventual much greater reward in an after-life. On the other hand, if the recipient is in a very vulnerable position, and the gift has much greater value for him or her than to the giver, and the giver intends to gain power through it, then it can slip further towards the theft end of the continuum; the giver is stealing the self-respect of the recipient. Similarly, theft can be a pure taking of something by force or, in the case of a 'con-man', a false reciprocity which amounts to the same thing. It can also involve receipt by the victim of the dubious kind of 'insurance' provided by protection rackets. But then there is a whole range of bad deals which involve an element of deception and false promise on the part of the seller and self-deception on the part of the buyer; these deals are not pure theft, but are certainly not fully reciprocal. Many social relationships have this characteristic, that one person takes more than he or she is prepared to give.

It is important to note at this point that there is no hard and fast distinction here between social and economic exchanges; both can take place anywhere on this continuum, and both involve the giving and receiving of valued possessions whether these be tangible goods measured in money equivalents or non-material goods such as one's good name. Furthermore, the marginal analysis of the economist can be used to explain social exchanges almost as well as it can exchanges in a recognised market (Blau, 1964), and, as Hayek admits, there are underlying social desires and expectations which fuel economic exchanges: 'Economic considerations are merely those by which we reconcile and adjust our different purposes, none of which, in the last resort are purely economic' (Hayek, 1960, p. 35). How then can we attempt to evaluate different types of exchange? One fruitful distinction is between those whose analysis starts from the individual and assumes a competitive urge to obtain personal rewards, and those who start from the group and assume a capacity to co-operate in reaching common goals. For various reasons which cannot be gone into here, the labels egoist and altruist which have dominated the debate since Spencer made much of the distinction a century ago are very misleading, and so we might call the two approaches competitive individualist and co-operative mutualist. Their views on exchange are obviously not the only ones which can be held, but they are important because they illuminate some fundamental issues concerning the relationship of the fraternal impulse to time and habit-formation. The ideal-types which they set up can be compared and contrasted, both in the way in which they neglect or incorporate the tendency to habitual behaviour, and from evidence about the way exchanges actually work in practice. The co-operative view provides probably the strongest specification for exchanges which both build fraternity and incorporate it into habit. The competitive individualist view provides a kind of null hypothesis with which to test out the strength of the co-operative theory.

AN INDIVIDUALISTIC VIEW OF EXCHANGE

Firstly, the competitive individualist's approach to exchange is drawn from the writings of market liberal or 'new right' theorists. They share in common an outlook which tends to deny the need for fraternity in an economic market; social exchanges are not dealt with at all, as people are simply assumed to be free to pursue their own preferences, but there is a tacit assumption that, as Spencer makes clear (1890), people tend to be altruistic only within the family, and egoistic, or at least rationally in pursuit of their own private ends, outside the family. They emphasise the primacy of market exchanges in which individuals deliberately suppress habits of solidarity to become as Hayek puts it 'impersonal means to their fellows' (1960, p. 37). In their view there are two ways of justifying a form of exchange: in terms of fairness and in terms of effectiveness in satisfying the desires of the participants. Both arguments rely on the effect of linear time on repeated exchanges. Firstly, they argue that a social sense of justice is not necessary because in a free market exchanges are always equal and therefore perfectly fair; outcomes are, in the jargon, 'Pareto-optimal'. This is not really true though; as Marx points out (Marx and Engels, 1968, p. 78), only occasionally can exchanges be truly equal because, in the individual exchange, supply and demand may not be in equilibrium, and price may be fluctuating above or below the optimum; the customer may be paying more or less than she should. The classical economist argues though that over time the fluctuations even out, and so everyone is engaged in a more or less equal series of transactions. Then there is a fall-back argument that, even if a market is not completely free, it still guarantees greater productive and distributive efficiency than any other type of economic exchange.

There are several objections to these arguments. Firstly, over time co-operation does build up between people, and so subverts the impersonal signals of the market. Cartels are set up by the producers to fix prices and wages, thus creating a kind of equilibrium which is permanently in their favour. Cartels of workers and consumers are less effective; they involve greater numbers of people who find it more difficult to co-ordinate their actions, and who are usually less able to withdraw from the market because the exchanges they make involve immediate subsistence. On the other hand, in a moderately competitive market where there is some choice, the consumers can still 'vote with their feet'. However, the tendency of the producers to combine with finance capital and to become multi-national also enhances the power of one party to market exchanges at the expense of the other. The need to plan large-scale production runs and far-flung distribution networks leads to a need to predict demand in the future, and so advertising often seeks not only to convey misleading information to the consumer, but to stabilise markets by manufacturing demand; when this happens, not only is the argument from fairness compromised, but the argument from efficiency or effectiveness becomes meaningless. Then there is the distorting effect of the past unfair exchanges which are still having an effect on present markets. In general this argument refers to the way in which land and the resources it contains have been acquired by force of trickery rather than by fair dealing. In particular it refers to the 'law of the first price' by which colonial powers fixed the price of raw materials which were at the time of little value to the seller, and then used military might to enforce it. All of these are ways in which time future and time past can have effects on the present, making economic exchanges far from equal; instead of oscillating around the mid-point of the continuum suggested above, they move towards the pole marked 'theft'.

Such arguments are reinforced if we take into account the effect of acquired habits. Buyers go to the same sellers not because they are cheapest but because they have always done so, and because over time 'gratitude' is built up which overlays transactions with moral sentiments. Sellers exploit the 'buying habits' of the public, and attempt to build up over time that 'consumer loyalty' which gives crucial marketing advantage to the retailer (and which can even be valued as an asset known as the 'goodwill' of the business). In fact, consumer habits are the subject of continual market research so that suppliers can take them into account. Habit may also be a larger factor than the competitive individualist expects, even in the behaviour of the producers and distributors of goods and services, making them less efficient than they think they are. For instance, when a Scottish community co-operative recently tried to find out why they were being surcharged for the cost of delivering milk which by the time it reached the shop was already going sour, they discovered that distribution was being channelled from the co-operative wholesaler via a local firm which never touched the milk, but was sub-contracting to a haulier. The wholesaler was passing the shop every evening, to deliver at a depot from which it was being sent back the next day. When challenged to explain themselves, the wholesaler replied, 'It was always done that way' (*Co-operative News*, 16.12.86).

Such habitual behaviour can of course be exploited by competitors, and sadly (but not surprisingly) the market share of such co-operatives is diminishing. Are habits more entrenched than we imagine, or is the above type of example only found in slow-moving markets? Is there a strong correlation between the pace of change and the extent of habitual behaviour? It is an interesting question for research, to what extent in a competitive market, where technology is continually being advanced, the 'habit of innovation' (as Young calls it) destroys every other kind of habit. On the other hand, if habit is as ubiquitous a force in society as we suspect, there really may be less reliance on competition than the theorists expect and more on tried and tested methods which have worked in the past. Hayek is one 'individualistic competitive' theorist who, while espousing a free market in which innovation is given free rein, yet acknowledges the power of habit in cultural evolution. He defines culture as 'a tradition of learnt rules of conduct which have never been invented, and whose functions the acting individuals usually do not understand' (1978, p. 5). The rules are followed because they work. However, the greatest rewards go to those who break the rules successfully; in his view habit simply acts as a kind of repository of successful methods which can be imitated until superseded by better ones.

It is difficult to tell just how entrenched habit is when viewing a relatively stable market situation. However, when observing rapid changes one notices that many innovations (which are in essence remarkably simple and when shared seem rather obvious) can nevertheless have profound effects on a market which had been thought to be optimal in efficiency, but which had in fact grown complacent with habitual thinking; for example, the recent application of the concept of the 'convenience store' has profoundly affected the way in which retailers now see the potential of the small shop, yet it is a simple theory which succeeds more by the reworking of old concepts into a new synthesis than by the invention of new ones (Birchall, 1987). Lastly, there is an even bigger question-mark against the efficiency of competitive markets. Habits of business practice are fairly superficial and amenable, under the stimulus of competition, to reasoned discovery and correction, but what about the deeper, underlying social habits which are so basic to the character of a society as to be almost impossible to bring to the surface? How do these affect economic performance? Are the Japanese given a competitive edge not by their habit of innovation but by habits of co-operation and loyalty to the group which may derive originally from a pre-capitalist society? Do the deeply entrenched British habits of class superiority and inferiority (of which the different treatment of blue- and white-collar workers on the 'shop floor' is merely one superficial example) militate against productivity, and so on?

A CO-OPERATIVE MUTUALIST POINT OF VIEW

Support for a view opposite to that of the competitive individualists comes from political theorists variously known as co-operative, communitarian socialist or anarchist collectivist. From this point of view, markets are in reality riddled with habitual affective ties and allegiances which, when they operate against the less powerful parties to exchanges, operate unfairly, but which can also generate a sense of obligation between buyer and seller which has built up over time. When this happens, the market becomes simply another aspect of human society in which Kropotkin's 'sense of justice growing to become a habit' (1915, p. 58) operates. In fact, even Hayek admits that, dangerous though affective relations are to the impersonal signals of the market, they do underpin people's willingness to obey the law of contract. Co-operators go much further, claiming that all exchanges, no matter how impersonal, have to be underpinned with expectations of honesty and fair dealing which are generated by Simmel's pervasive 'sense of gratitude'; otherwise the market simply falls apart. Co-operators, following Kropotkin, picture an ideal exchange not as the immediate exchange of equivalents, abstracted from the flow of time, but as one in a chain of mutual aid transactions whose true character can only be assessed over time and in the light of deeply entrenched habits of reciprocal rights and obligations. This means that, when they take place within a mutualist context, it is misleading to try to place exchanges on the continuum between theft and the pure gift without knowing the full social context in which they occur.

Take, for instance, the following examples. Deep in suburbia it is lawn-manicuring time, and Mrs A lends Mr B a garden roller. An act of pure altruism one might think, but the next day Mr B pops over and mends Mrs A's garden gate. He has in a rough and ready way repaid her gift. But perhaps he does not get an opportunity to do so. Is he then in debt? He may be, but he may have given her some lawn seed the year before and so she may have been repaying a previous kindness, also in kind. This is in technical terms, serial, bilateral reciprocity. Another example. High up in the mountains, a storm is brewing, and Mr A enters a wooden cabin and helps himself to a meal from the larder, settles down to sleep for the night, and the next day moves on. Is this theft? In one context it is; he is a vagrant in a Swiss ski-resort, and an efficient police force track him down and charge him with breaking and entering and stealing a can of baked beans. In another context he is a fur trapper in the Canadian wilderness, and considers it his right to accept another person's hospitality even though the other person is not there. Meanwhile Mr B the owner of the hut is sleeping in the hut of Mr C who may at some time during the season end up eating out of Mr A's larder; this is serial, multi-lateral reciprocity. It presupposes a network of individuals or groups with sufficient faith in each other's goodwill to act on very general norms of what to the outsider would seem like generosity but is in fact mutual aid. Unlike the pure altruistic gift and market exchange, it cannot be understood outside networks that extend in both space and time. It is like a child's dot-to-dot picture which can only be recognised when all the dots are joined up. Another example. A family have just moved into a farm in Wales. It is high summer and the grass is ready for cutting. A tractor pulls into the field and several people jump out. Together they harvest the hay and, after it has dried, come back to gather it into the barn, all without needing to discuss the business in hand. This is a different kind of reciprocity which we might call serial collective.

The last example raises two important questions. Firstly, is the new family expected to join in with the harvest on other people's fields? If so, how many? It may be that so much goodwill has built up over time that if this family 'freeload' on the others it does not matter. On the other hand, if the exchangers are uncertain about the long-term future of their relationships, they may wish to assess the value of the exchange and come to an agreement which is more barter than mutual aid; an exchange of equivalents which Gouldner calls 'heteromorphic' (1960), but which is essentially a market exchange of items of equivalent value. In this case freeloaders are definitely seen as a threat to the survival of the exchange relationship; in a study of newly formed housing co-ops, acute anxiety was found among members concerning the effects of freeloaders on the active co-operators (Birchall, 1988, ch. 7). Insecure co-operators may even wish to build up 'social capital' (a direct reference to the market-based thinking which can underlie even social relationships when they are uncertain) by giving more than they receive. When they are secure then, exchange relationships can depart from strict exchange of equivalents and even from strict concern with the value of what is being exchanged; over time (perhaps even over generations) things even out. When they are insecure such relationships approximate more to market values and look to short-term payoffs.

Secondly, the same example suggests that, in reaching back

to a practice which has largely been superseded in Britain (the coming of the silage method of grass-storage more or less killing it off recently in Wales), collective mutual exchanges are almost impossible to maintain in modern market-dominate societies. In fact, Kropotkin's examples of mutual aid mainly come from pre-industrial societies, and their modern equivalents (he cites the consumer co-operative movement and various voluntary associations) are far from being the central institutions of society. Herbert Spencer has put the problem succinctly. He lists four types of co-operation:

1. Like efforts for like ends, simultaneously enjoyed.
2. Like efforts for like ends, not simultaneously enjoyed.
3. Unlike efforts for like ends.
4. Unlike efforts for unlike ends (Spencer, 1980, p. 146).

The first three involve various types of mutual aid: from the primitive hunting band of the first type, through the collective co-operation practised in settled agricultural societies of the second type (in which our last example fits), to a more sophisticated form of exchange in which the division of labour appears, typified by the worker or consumer co-operative. The fourth type signals the arrival of the 'individualistic competitive' free market, and the key point is that in Spencer's view the fourth form of exchange has superseded the first three.

Whether one agrees with Spencer or with Kropotkin depends to a large extent on whether one takes evidence from the economic sphere or from the social. There is, for example, plenty of evidence to suggest that mutual aid is still a common form of exchange between neighbours (Bulmer, 1986) and in voluntary organisations, but money markets seem, despite a flourishing 'informal' economy, to dominate in the economic sphere. How would a co-operator explain this undermining and supersession of habits of mutual aid which are after all those which have been practised in the economic sphere by most people in most societies until the recent dramatic spread of the market economy? Kropotkin blames the rise of the modern centralised state which in his view has undermined mutuality by destroying the old self-regulating institutions of simpler societies. However, the state may itself be a response to a problem of social order caused by deeper economic changes which we could describe (while running the risk of over-simplification) as the move from mutualistic to individualistic exchange.

TIME, HABIT AND THE DEVELOPMENT OF CO-OPERATION

There are two approaches which suggest that co-operation is the normal outcome of exchanges over time, and which also confront the problem of why the 'abnormal' individualistic form has largely taken over in the economic sphere. The first is Axelrod's work on games theory, the second Durkheim's on social solidarity. Robert Axelrod posed the question which is crucial to a mutual view of exchange; 'When should a person co-operate, and when should a person be selfish, in an on-going interaction with another person?' (1984, p. vii). To find out, he invited games theorists to submit strategies for a simple game known as the 'iterated Prisoner's Dilemma'. Two players interact by choosing to co-operate or to defect. If both co-operate they score two points each, if one co-operates and the other defects the defector scores three points and the co-operator one, and if both defect they each score one point. Using a computer to run each strategy with every other, Axelrod found that a highly co-operative strategy called 'Tit for Tat' won every time. This strategy could hardly have been more simple; it was simply to begin by co-operating, and thereafter always to do what the other player did on the previous move. The game assumed that people are self-interested, but strategies based on varying amounts of defection failed because in the long run they produced a poorer outcome for all. Under the game's conditions, in the long run co-operation maximised the payoffs for both egoists.

These conditions include, firstly, that the interaction continues for an indefinite number of times; no one knows when it will end. Secondly, each player is assumed to recognise the other and to remember the history of the interaction. Thirdly, players cannot enforce their will on, eliminate or run away from the other, nor can they change the other person's payoffs. Fourthly, the future has to be taken into account in that the players may have to meet again. The payoffs need not be comparable or symmetric; they only have to be measured relative to each other. This means that the strategy covers looser forms of mutual aid as well as strict market exchange of equivalents. Nor need the choices be entirely deliberative; as Axelrod says, 'Their strategies may simply reflect standard operating procedures, rules of thumb, instincts, *habits*, or imitation' (1984, p. 18, my emphasis). The robustness of the co-operative strategy is illustrated by the fact that in the First World War it grew up between antagonists who during repeated exchanges found a way of tacitly agreeing to limit the damage done to each other. In the process they developed feelings of fraternity which altered the payoffs to make mutual restraint even more valuable than it had been before. In summary, then, exchanges over time tend to turn into co-operative or mutual exchanges based on a mixture of memory and habit, and producing feelings of fraternity which make the exchanges even more valued. Time, habit and the fraternal impulse really are a powerful combination.

Why, then, is there not more co-operation in the real world? One answer is that there is far more than we think, both in particular and in general; Axelrod cites evidence of extensive mutual aid in the supposedly adversarial setting of the American Senate, and Young emphasises the co-operative nature of communications and information-sharing on a world-wide scale (1988, ch. 5 on 'Social Evolution'). But another answer might be that often the conditions are not right, that in some way real interactions do not conform to the rules of the game. If people interact for only short periods of time, they never build up stable relationships in which each knows what to expect from the other. Or if, as the soldiers did in the trenches, they interact for long periods but third parties (in this case the officers, who organised randomly timed raids) constantly intervene to change the rules and make the outcomes unpredictable, co-operation will be inhibited or destroyed.

Confirmation of these points comes from Durkheim. Like other co-operators he believes that fraternity or social solidarity is central to the problem of social order; it 'is the least dispensable, the strictly necessary, the daily bread without which societies cannot exist' (1964, p. 51). But he goes one stage further in identifying not social interaction or exchange as the key to the generation of fraternity, but the division of labour. However, the difference is slight, since it is the division or specialisation of functions among individuals which produces a need for repeated exchanges. Thus 'The ideal of human fraternity can be realised only in proportion to the progress of the division of labour' (Durkheim, 1964, p. 406). Like Axelrod, he sees co-operation as emerging automatically as a result of the need to exchange in a sustained way over time, but then he also has to bring in the exceptions to the rule. Firstly, he confirms the importance of stability of relationships over time; the 'anomic' division of labour occurs when economic differentiation outstrips the development of the new norms which are needed to regulate it. Or to put it another way, a too-fast pace of economic change destroys customary habits of thinking and acting, and does not allow for the growth of new ones. Secondly, he confirms the danger from third parties who break into the exchange to change its rules and render its outcomes uncertain; the 'forced' division of labour results, in which new norms or customs do not grow up naturally but are imposed; the rules of the game are changed to advantage the powerful in society against the weak, and a 'troubled and imperfect solidarity' results.

CONCLUSION

There are at least two conditions, then, for the growth of the fraternal impulse: stability of social interactions over time and adherence to some basic rules which ensure that the exchanges are, as Durkheim calls them, 'spontaneous' rather than forced. When the conditions are right, habits build up over time and fix the exchanges into enduring social bonds. But when the conditions are wrong, the generation of the fraternal impulse can be inhibited, and social order becomes problematic. We might ask why societies which seem to be torn by political, religious or racial conflict survive at all? Marx reached the conclusion that at a certain, quite predictable point, the 'competitive individualist' society would destroy itself, but he was wrong. The answer may be found in the nature of habit itself. As Young has pointed out (1988), the power of habit can be used against people to habituate them to essentially immoral exchanges in which they co-operate but always lose; habit is a two-edged sword.

This can be seen more clearly if we turn from the narrow focus of individual interactions and exchange theory to the wider approach hinted at earlier, which emphasises that individuals are socialised into a common culture. We are all born into a socially constructed everyday world which we take for granted, and which is a product of incalculable millions of social exchanges which have produced common habits and associated feelings. Language, skills, gender roles, beliefs, all are absorbed long before any critical faculties arise to question them. From this aspect, the competitive individualist plays on the surface of things, while the deep structure of co-operation provides the supporting framework. It is because there is so much taken-for-granted similarity between people that they can concentrate on the differences, assert their individuality. The deep fraternity does not necessarily foster fraternal exchanges, but may merely keep the competitive spirit in check within certain commonly accepted ground-rules. However, there is still a problem of social order. Fraternity is a deep reservoir of common culture, which can be drawn on to replenish the surface goodwill between people which is lost when they compete unfairly so as to fix the outcome of social and economic exchanges. The reservoir is as wide as society and as deep as the centuries, but unless it is continually replenished from somewhere, it will gradually dry up. While the competitive individualists are learning how to examine the habits of the rest in order to exploit them, there is always the possibility of a counter-tendency, that the co-operators examine their habits too, and in the end refuse to co-operate. The survival of the fittest can only be assured when the whole species co-operates to support them. If that co-operation is consciously withdrawn, the reservoir will dry up, and the survival of the whole species may be endangered.

REFERENCES

Axelrod, R. (1984), *The Evolution of Co-operation*, New York, Basic Books.

Berger, P. and Luckmann, T. (1971), *The Social Construction of Reality*, Harmondsworth, Penguin.

Birchall, J. (1987), *Save our Shop: the Fall and Rise of the Small Co-operative Store*, Manchester, Holyoake Press.

Birchall J. (1988), *Building Communities: the Co-operative Way*, London, Routledge & Kegan Paul.

Blau, P. (1964), *Exchange and Power in Social Life*, London, John Wiley.

Bulmer, M. (1986), *Neighbours: the Work of Philip Abrams*, Cambridge, Cambridge University Press.

Camic, C. (1986), 'The matter of habit', in *American Journal of Sociology*, vol. 19, no. 5, (March 1986) pp. 1039–87.

Cohen, P.S. (1968), *Modern Social Theory*, London, Heinemann.

Durkheim, E. (1964), *The Division of Labour in Society*, New York, Free Press.

Giddens, A. (1979), *Central Problems in Social Theory*, Berkeley and Los Angeles, University of California Press.

Gouldner, A.W. (1960), 'The norm of reciprocity, a preliminary statement', in *American Sociological Review*, vol. 25, no. 2, pp. 166–78.

Halsey, A.H. (1978), *Change in British Society*, Oxford, Oxford University Press.

Hayek, F.A. (1960), *The Constitution of Liberty*, London, Routledge & Kegan Paul.

Hayek, F.A. (1978), *The Three Sources of Human Values*, London, London School of Economics.

Homans, G.C. (1974), *Social Behaviour, its Elementary Forms*, New York, Harcourt Brace Jovanovitch.

Kropotkin, P. (1915), *Mutual Aid: A Factor of Evolution*, Boston Massachusetts, Extending Horizons.

Kuhn, T.S. (1962), *The Structure of Scientific Revolutions*, Chicago, University of Chicago Press.

Marshall, T.H. (1977), *Class, Citizenship and Social Development*, Chicago, University of Chicago Press.

Marx, K. and Engels, F. (1968), *Selected Works*, London, Lawrence & Wishart.

Pennock, J.R. (1950), *Liberal Democracy, its Merits and Prospects*, New York, Rinehart.

Perry, R.B. (1944), *Puritanism and Democracy*, New York, Vanguard Press.

Rawls, J. (1973), *A Theory of Justice*, Oxford, Oxford University Press.
Simmel, G. (1964), *The Sociology of Georg Simmel*, New York, Free Press.
Sorokin, P.A. (1943), 'Socio-cultural time', in *Socio-Cultural Causality, Space, Time*, Durham N.C., Duke University Press.
Spencer, H. (1890), *The Data of Ethics*, London, Williams & Norgate.
Tawney, R.H. (1964), *Equality*, London, George Allen & Unwin.
Young, M. (1988), *The Metronomic Society*, Harvard University Press.

CHAPTER 9

SOCIAL VERSUS NATURAL TIME, A TRADITIONAL DISTINCTION RE-EXAMINED[1]

Barbara Adam

INTRODUCTION

The study of time destroys the sanctity of disciplinary boundaries. Sociologists have to become philosophers and psychologists. Physicists find that they need to include their own consciousness in the understanding of the temporalities of their subject matter, and some of their writings extend to biology and psychology. Interdisciplinary collections of writings on time, however, show that the bridges have not yet been joined.[2] Yet there is a need for it, since social scientists, for example, constantly encounter physical and biological time as an integral aspect of the social lives they study. Just as sociologists cannot evade the physical and biological temporalities of everyday reality, so physicists cannot ignore that not only their own consciousness, but also their social context, has a bearing on the reality they study.

It is suggested in this chapter that social scientists[3] must get to know physical and biological time. They must not only learn to recognise natural time so that they may safely exclude it from their social science analysis, but need to get to know natural time in its own right and in its relation to those aspects that are conventionally designated 'social time'. This is not an easy task. The reason for the difficulty, however, does not lie in the subject matter, i.e. physical and biological time, but relates to the social scientists' understanding of the nature of the natural sciences. This is a nineteenth-century understanding of science that does not take account of the developments and the fundamental conceptual shifts that have taken place during this century. It is these conceptual changes and their bearing on social science that I am concerned with here.

The nineteenth-century conceptualisation of natural science is reflected in the social sciences' understanding of physical and biological time. In this chapter I therefore look at some classical social science conceptualisations of nature and the concomitant views of natural time. I shall then explore these in relation to some of the contemporary natural scientists' understandings of the nature and the temporalities of their subject matters.[4] I am suggesting that it is not just important to understand those times in their relation, but that a contemporary understanding of natural time is a pre-requisite for an understanding of social life that seeks to take explicit account of time.

TRADITIONAL DISTINCTIONS

I am not here concerned with the dualism of synchronic and diachronic analysis in social theory, or the need for sociology to embrace the historical dimension. Giddens (1979, 1981) and Martins (1974), for example, present these debates and show that time is conventionally associated with theories of change, whilst synchronic analysis is charged with excluding the dimension of time. Apart from change, time is generally understood to have entered classical sociology in a number of other ways. Action theory, for example, is concerned with purposive behaviour and is thus by definition future-orientated. In Schutz's work purposive action is further shown to be simultaneously future-orientated, reflective and continuous. Self, identity, the other, biography and career, key concepts in symbolic interactionism, are all fundamentally 'time loaded'. So too are historical and evolutionary approaches. The latter postulate immanent and endogenously determined change, whilst the former include time, regardless of whether they focus on chronologically ordered data, events, long-term cumulative processes or discontinuous episodes. Both systems theory and Marxist analyses make use of those latter aspects to varying degrees. Marxist analyses, however, are shown by Giddens (1981) and Heirich (1964) to include the further aspect of time as a resource; as something that may be used or sold and exchanged as a commodity on the labour market. Lastly, it must not be forgotten that time also enters as measure in social science studies; that it is a key element of quantitative research.

This rather cursory list offers a sketch of the range of ways in which time has traditionally been understood to have entered social science. Common to all these approaches is their treatment of time as incidental and unproblematic. Time here, as Lewis and Weigert (1981) and Lüscher (1974) point out, is neither investigated in its own right nor included as an independent variable. Instead of being the centre of social theory it is, Giddens (1979, p. 202) suggests, treated more as an environment or a parameter within which social life is enacted. Where time has received some secondary attention, it has been understood primarily through one aspect only, be it as change, as past, present and future extension, as resource, or as measure. Hägerstrand (1985) distinguishes between embedded time and symbolic time, between the time *in* events, things, and conditions, and the conceptualised time of clocks and calendars. In its embedded form, he proposes, time is an integral part of the social scientists' tacit working knowledge. In its symbolised form, on the other hand, clock-time predominates.

I want to concentrate on those approaches that focus on time *explicitly* in both its symbolised and embedded form. Once time comes to be understood as historically and culturally relative, the symbolised form of clocks and calendars ceases to be all-pervasive. Both historical and anthropological studies of the concept of time and its relation to social structure, relations of production, technologies and cosmologies, have had such a relativising influence.[5]

When time comes to be focussed on in its own right, one begins to become aware of what seems to be an infinite complexity of 'times'; a myriad of aspects, dimensions, and meanings. Pandora's box has been opened. In the myth, the box when opened released evil and retained only hope. In similar fashion a focus on time creates conceptual chaos and confusion while, at the same time, holding the promise of new order. From being an unproblematic aspect of diachronic analysis and quantitative research, time suddenly comes to be understood as centrally implicated in the conditions of existence, in experience, action, interaction, exchange, language, self-understanding, historicity, in information, institutions, social organisation, technology, and even the power of the state; each containing their own multitude of times within.

Social scientists, threatened with drowning in this 'flood of times', needed to engage in the process of categorisation and redrawing of the boundaries of the nature of 'the social'. It is the assumptions entailed in this redefinition and delineating of the subject matter of social science in relation to that of natural science that I am centrally concerned with in this paper.

For the purpose of this argument I want to focus on just three central and traditionally accepted ideas of social science. All three are concerned with the distinction between society and nature[6] and, either explicitly or by default, with defining social time in contradistinction to natural time. The latter time distinctions appear to have been formulated in response to the theoretical uncertainties created by an explicit focus on time.

The first set of ideas contains the assumptions that nature is quantifiable, and that any understanding of human social life needs to take account of quality. Whilst natural science is thought to deal with invariant relations, universal connections, and laws that hold beyond space and time, social science is argued to be additionally concerned with non-mechanical relations and aspects that are fundamentally modifiable by knowledge in conjunction with the conditions of existence. Social science is further understood to deal with mediated knowledge and experiences in which both content and context matter. This qualitative aspect then is seen as an addition to quantity, and conceptualised as a prerogative of human social life.

With respect to time, we encounter a parallel set of ideas. Ever since Sorokin and Merton's (1937) seminal paper, social time has been contrasted with natural time, and clock-time equated with the homogeneous time of the physical world of diurnal and seasonal cycles. Writing about this distinction, Sorokin and Merton suggest,

> Astronomical time is uniform, homogeneous; it is purely quantitative, shorn of qualitative variations. Can we so characterize social time? Obviously not – there are holidays, days devoted to the observance of particular civil functions, 'lucky' and 'unlucky' days, market days, etc. (Sorokin and Merton, 1937, p. 621)

From the publication of this paper onwards, physical time has become understood to come in fixed and infinitely divisible units that may be measured. Human time, on the other hand, may include the latter but cannot be fully explained by it. The experience, the content and the social frames of reference, it is argued by Lewis and Weigert (1981), are fundamentally qualitative.

Clock-time then, is thought to be an adequate tool for the natural sciences to measure their subject matter, whilst there are time aspects of social life that cannot be quantified. One can neither measure the experience, awareness, consciousness, or knowledge of time, for example, nor can one quantify identity, action towards projects, or the ordering of activities according to priorities. Whilst people may use clock-time to co-ordinate, synchronise and order their social life, this clock-time is the means by which such a social life is achieved, but not an explanation for it. Nature, on the other hand, is generally thought to lack such qualitative time aspects, to be far less complicated and thus to be amenable to quantitative measure.

Underlying those thoughts is the philosopher Bergson's distinction between 'temps' and 'durée', where 'temps' is characterised as mathematical, even flowing, and divisible into spatial units. This 'temps' was the classical fomulation of quantitative time, and it seems to be this characterisation that is being drawn on by social scientists who have become aware of the limitations of clock-time for the understanding of social life.[7] Because clock-time is also such a central tool for all quantitative science, I shall be giving more attention to the nature of this particular type of time than to some of the others.

In a second, quite closely related, idea it is suggested that nature may be understood objectively. Natural scientists, explains Giddens (1976), stand in a subject–object relationship to their subject matter. In distinction to social scientists, they may study objects directly and apply a causal framework of analysis. Such direct causal links are, however, no longer thought to pertain to human society, where that which is studied is an unintended outcome of intended actions. Nature is understood to be predictable because its regularities – be they causal, statistical or probabilistic – are timeless. Human societies, on the other hand, are organised around values, goals, morals, ethics and hopes, whilst simultaneously being influenced by tradition, habits and legitimised meanings. Human societies, it is argued, are fundamentally historical.

If we now focus more explicitly on time, we again find a parallel set of criteria by which social time is being distinguished. Nature, suggest Elias (1982a, 1982b and 1984) and Jaques (1982), may be grasped through the objective, absolute relations of earlier and later, whilst an understanding of human society must necessarily include the subjective, relative, experiential aspects of past, present, and future. Because it is presumed firstly that the physical world is subject to laws, and secondly that natural processes entail no element of choice, purpose, or meaning, the future and past are thought to be irrelevant. The experientially based relations of past and future, it is argued, are inappropriately used if they are applied to causal event chains in nature. Human beings, on the other hand, are thought not only to live and die like any other creature, but to experience themselves and their social life as extending into the past and future. Their birth, growth, ageing and death can be located in an objective time, whilst their experience of identity, their knowledge and motives fundamentally depend on the subjective standpoint for their meaning. Planning of future actions is influenced by the past, and motives are thought to have either a future or a past bias depending on whether actions are orientated towards projects or being rationalised retrospectively.[8]

This particular distinction between understanding events as having happened earlier or later or, alternatively, in terms of their being past, present, or future, dates back to the turn of this century when the philosopher McTaggart (1927) formulated the distinction as the A- and B-series of time. He used his conceptualisation for a logical argument against the reality of time. Social theorists have not been interested in that specific aspect of the distinction, but appropriated it as a useful tool for social analysis.

In brief, McTaggart argues that events can be understood as 'in time' where the relation between them is defined in a permanent and absolute way. If event A happened before event X, then A will always be earlier then event X. In other words, Caesar died before Queen Victoria, people are young before they grow old, and the shiny state of the car precedes the rusty one. These are taken to be 'timelessly true' statements of temporal relations. McTaggart calls this 'tenseless' conception of relations between events the B-series of time. The same events, however, can also be understood with regard to their past-ness, present-ness and futurity. As such, they are understood in relative terms because, he argues, to define something as past, present, or future depends on an observer and on the surrounding relations. Tensed statements are context-dependent, and this makes them inherently relative, impermanent and temporal. He calls these the A-series of time. He understands this A-series to be more fundamental, not only because it incorporates the reality of change, but also because relations of earlier and later are reducible to the A-series, but not vice versa.

Out of this classical discussion grew a modern conceptualisation where relations of earlier and later are understood as absolute, objective, and time-less, whilst past, present, and future are deemed subjective, relative, temporal and essential for any expression of change. Elias (1982a, 1982b and 1984) designates them experiential and structural relations respectively, whilst Jaques (1982) distinguishes between the temporal axes of intent and succession. Regardless of their differences with respect to the detail of those distinctions, social theorists agree that the B-series, the objective structural relations of succession, suffices for the description of the non-human world. Luhmann's (1978) insistence that 'there cannot exist a single state for which the past and the future would be irrelevant' (p. 99) is an isolated exception here.

The third characteristic attributed to social time is also the most widely accepted in sociology. It simply states that all time is social time.[9] Time, argues Lauer (1981, p. vii), does not exist independent of human life but is always socially constructed. This idea appears in many guises, but one of its key arguments is that time is always dependent on the meaning of time. Time presupposes a view of time, irrespective of the symbolic form it may take in different cultures of the world. Time as a symbolic expression is thought to be always collectively formulated. Regardless of whether it is used for understanding physical phenomena, natural processes of ageing, or social processes of time control, the meaning is always arrived at socially. The general parallel to this idea is that the world is only as formulated; there is no reality outside language.

Other variants are closely linked to this central idea that time is a socially constructed symbol through which we explain the world. Only humans, argues Moore (1963), use time to control, regulate, order or synchronise their social life, and it is only in human societies that the members try to cheat, structure, make, or measure time. Only humans contemplate and study their own creations. Only they may become aware that time may be in events, but that only people lead an 'in time' existence; an existence in which the human creation is imposed as an external frame of reference within which social life is enacted.

Where social scientists sought to redefine the boundaries of their subject matter, and distinguish social time from natural time they have, with few exceptions,[10] drawn on classical philosophy for their characterisation of natural time. They have appropriated Bergson's division into 'temps' and 'durée', and McTaggart's classification into an A- and B-series of time for that purpose.[11] In agreement with Elias (1984, pp. xv–xvi), however, I hold that a comprehensive understanding of time is not possible from a position where nature and society are treated as separate. Not humans and nature, but human society *as* nature, is the basis from which to understand the multiplicity of times in their relation and syntheses.

I want to argue further that the understanding outlined above is based on an outdated model of the natural sciences and with it an inadequate understanding of natural time. This ought to be rectified since our conceptualisations of the natural sciences, nature, and natural time affect our understanding of social time and social life.

I propose to use the conceptual chaos that is created by an explicit focus on time as an opportunity to do something different; to let go of the habit of turning to classical philosophy for our categorisations, and to turn instead to the natural sciences for finding out about natural time. I shall explore how some contemporary physicists and biologists understand the inherent time aspects of the realities they study, and what the implications might be for our understanding of social life.[12] In this chapter, however, I am not concerned to familiarise the reader with the laws and theories of contemporary physics and biology, but seek to extract the principles of time they entail, in order to establish a relation between these and the aspects of time central to an understanding of contemporary social life.

NATURAL TIME

Because twentieth-century physicists could no longer contain their findings and emergent understandings within the classical nineteenth-century Newtonian physics, they had to become philosophers in order to conceptualise the strange reality that was forced upon them through their discoveries. It is this aspect of their work which is accessible to interested lay persons and therefore allows an insight into an otherwise closed scientific world. Where physicists have thus been concerned with the foundations and base assumptions of their knowledge, they have produced work which is quite explicitly concerned with the nature of time, space, light or matter: those otherwise most taken-for-granted aspects of their work.[13]

A focus on time in physics reveals three clusters of approaches that have a direct relevance to social theory.[14] The first includes time as both a measure and as a quantity to be measured. As such, it is used in Newtonian laws, mechanics, and the theory of relativity, for example. The second is concerned with directional processes and events as expressed in the laws of thermodynamics or the theory of dissipative structures. Quantum theory provides the third approach to physical time, having found the ultimate reality to be fundamentally temporal. Thus, there follows a brief exploration of the Newtonian measure, the thermodynamic change, and quantum temporality.

When we use time in social science it is likely to be Newtonian time. When we locate events in the time grids of clocks and calendars, when we make statements about duration or movements from A to B, when we calculate rates of change, and when we study the sequencing and timing of social action, we do in fact make use of Newtonian time. As an integral aspect of the working knowledge of everyday life it is, however, not explored in its own right; its meaning never thought about. Because it is the most utilised time of social science – in both theory and empirical research – it seems particularly important to make the nature of Newtonian time explicit.

In Newtonian physics time is conceptualised as a unit equal to the other fundamental quantities of mass and space and it is in this form that it enters Newton's laws. Motion cannot be described without time. If something moves, then it changes its position in space in a certain time interval. This classical, physical time 't' is conceived as an invariant quantity, infinitely divisible into space-like units, expressible as number, and measurable in length. It is time taken – the duration between events – which is unaffected by the transformation it describes. There is change, but this is external and related to position, rather than changing that which does the moving. It is one of this 't' time's major characteristics that it can only distinguish between states on a before and after basis, but not recognise pasts and futures: the very thing that social scientists insist is central to their analyses.

Motion, in theoretical physics, can proceed forward or backwards and is invariant with respect to time. In the Newtonian equations, it can thus be argued, motion proceeds in 'non-temporal time'. What do the physicists mean by this idea of non-temporal time'? Physicists use this term for the 't'-coordinate because it excludes all those aspects which form an integral part of human time such as the importance of past, present or future, rhythmic repetition with variation, processes of interaction and their irreversible outcomes, and change. The only change which can be expressed in mathematical terms is acceleration; in other words, the change in the state of motion. But again, this equation is indifferent to any distinction between the past or the future and has no need for any location in a present unique now.

Newton conceived this physical 't'-coordinate to be located in an absolute time. He believed all things and events to have a distinct position in space and to occur at unique moments in time. He thought that

> Absolute, true and mathematical time, of itself, and from its own nature, flows equably without relation to anything external, and by another name is called duration. . . . All motions may be accelerated or retarded, but the flowing of absolute time is not liable to change. (quoted in Shallis 1983, p. 17)

All Newton's measurements and laws pertained to the motion of things 'in time' only, but not to his postulated absolute time. His mathematical technique was applied to calculate rates of change. It did not deal with change explicitly as an external quantity, but incorporates time only implicitly in the description of events. Absolute time, which he conceived of as flowing continually without relation to anything, has been removed from the description. Motion has been stabilised. We can thus conclude that Newton's mathematical description of changing events is dealing with a non-temporal quantity which is generally applicable without the need for a temporal context, or reference to either the past, present or future.

At this point it may be useful to draw together some of those issues which are important to remember. These help to make explicit what is entailed when time is conceptualised as a measure of motion, duration, and rates of change. In Newtonian physics the search is for the absolute and the eternal. The process by which this is achieved is one of extracting single units from interactive wholes, and seeking that unit-part in the ultimate simplicity. Once one has penetrated to this level of simplicity, the motion of atoms is thought to be time-reversible. Complexity, on the other hand, by being understood as the essence of the macro world, is held to be responsible for irreversible processes and unidirectionality. This time-reversibility of the micro world is postulated on the basis of idealisations. If one excludes friction, if one excludes gravity, if one excludes radiation and electro-magnetism, if one excludes interaction and complexity, then one is left with a universe of perfect symmetry, single parts in motion, and non-temporal, directionless time. Time is further defined operationally, linked to number and the measure of motion, duration and rate. As such it depends on the postulation of an absolute time within which the laws of motion find their expression. This absolute time, however, has never been successfully incorporated; it has never become an integral part of Newtonian time.

What about clock-time? Could that not be understood as representing Newton's absolute and flowing time? Shallis (1983, pp. 63–9) thinks not, because the clock is only used to count regular and periodic intervals of frequency. The direction of clock-time is irrelevant for the measurement of the units, since the latter are conceptualised as consistent in either direction. The directionality of the clock, I would like to suggest, is given by the convention of our number system. The idea that Newton's time is that of the clock is thus a complex and problematic one since Newton's laws and the motion of the oscillations in a clock must be conceptualised as in principle reversible. That which makes the clock-time irreversible, on the other hand, is the direction of the number system which determines that the number two follows the one and precedes the three. It is this which gives direction to the recorded oscillations and not the motion, since the latter is understood as invariant with respect to time, able to move forward or backwards in (absolute) time. Whilst the Newtonian 't'-time, entailed in the clocks, is in principle reversible, clocks as machines are designed to have their parts interact sequentially and are as such an integral and coherent expression of the causally orientated Newtonian science.

Clocks can be seen as mechanical models of the universe, with their clockface representing time visually, as distance travelled in space. They not only entail time as measure, but they 'measure time'. This duality of time in clocks, as absolute measure and as tool to measure time, needs brief attention. The aim of the measure is invariance and precision, whilst the very essence of that which is being measured is repetition with variance. No two seasons are ever alike, nor do the configurations of the planets repeat themselves. The earth's rotation too is slightly irregular, relative to other time-keepers. All natural time sources are characterised by repetition with variation, and clocks and calendars, our devices to record and measure natural time, have to be calibrated with those natural sources. This always involves the human device being adjusted to nature's time. The chronological time concept that Sorokin and Merton (1937) designated physical time is thus neither an aspect of nature nor a characteristic of the 't'-coordinate. It is exclusively a human creation.

As a machine the clock further incorporates the time aspects of mechanical organisation. Like any other mechanical device, it is designed so that its parts interact in a certain sequence, at the appropriate speed, at the correct intervals. All interactions have to be accurately timed and synchronised, within whole cycles of interaction. Lastly, all internal actions as well as the whole cycles are designed for repetition without variance.

Thus, my findings suggest the clock to be a material expression of Newtonian mechanics and as a machine to entail time as both the spatial measure of motion and an aspect of mechanical organisation, i.e. duration, timing, sequence, rate and periodicity (repetition of completed cycles of change). The direction of time was found to matter with respect to the number system, but not with regard to the oscillations that are being counted. The meaning of clock-time, on the other hand, is contained neither in the number system, nor in the oscillations, but depends on the context and real environmental changes, since the clock cannot tell us whether it is 3 o'clock in the afternoon or the morning, summer or winter, in the Southern or Northern hemisphere. With respect to the A- and B-series of time there is an interesting twist here. Clock-time is meant to be an expression of the absolute, objective B-series of structural relations of before and after, or earlier and later. Yet the clock always defines a now-point as significant, since earlier temporal points are simultaneously those that will follow. In other words, if it is 3 o'clock now, then 9 o'clock is both earlier and later.

The time entailed in clocks, I suggest, must be appreciated as an idea in practice: as a material expression of a particular understanding of the natural world, in which time is conceptualised through motion without change, as a spatial quantity which is infinitely divisible into units, which can be numerically defined, and where parts interact within abstracted wholes, to a design of invariant repetition, accurate timing, sequencing and spacing.

This Newtonian clock-time conceptualisation has deeply penetrated the everyday life of modern, industrialised society. Interacting with any machine – driving a car, for example – necessitates a working knowledge of those principles. The Newtonian time of both the measure and the mechanical organisation forms an integral aspect of our contemporary life. What is far more doubtful is its applicability to nonhuman nature. Yet, in recognising the limitations of clock-time for understanding social life, social theorists have designated it physical time. Whilst they thought it problematic for their own use, they approved it as an appropriate tool for the natural sciences.

Having gone into some depth on the nature of the Newtonian 't'-coordinate and clock-time, I shall present the rest of natural time in a much less detailed form. Rather than going through single steps of the 'evidence', I shall present an overview of the results of my research on natural time. I shall then use these as a base from which to question the idea that nature may be grasped, described and explained through the structural relations of the B-series of time.

Einstein retained time as a measure of motion in his theories of relativity. His approach to time was thus quite closely wedded to that of classical physics. Yet, by defining time as local and contextual, and by postulating its source in the acceleration of motion, he changed the meaning of time as measure of motion. Time, he argued, is relative to the framework of the observer, and so is the definition of an event as having happened before or after. Despite the fact that Einstein dealt with acceleration speeds at or near the speed of light, the principle of his argument applies generally and thus affects the objective status of the B-series of time. On the basis of Einstein's theories, time ought to be conceptualised as both relative and internal to the system of observation.

Due to Einstein's work, the close connection between time, light and causality has been made more explicit and both Jones (1983) and Shallis (1983) stress the intimate and mysterious connection between light, causality and our understanding of time. Light, they argue, is not only a prerequisite for the communication and transmission of signals but also presents the limit to physically possible, causal connections. Causal relationships, according to Jones (1983, p. 112), can only hold between events if these are separated temporally so that a light signal, or anything slower, can pass between them. Causality can only be envisaged because it takes time to cross distance. The B-relations of time are, of course, an integral aspect of a causally connected world; cause and effect are just a different way of expressing relations of 'before' and 'after'.

For quantum theorists, however, those causal relations of succession are mostly irrelevant. At the sub-atomic level the physical world emerges as an indivisible, dynamic patterned whole. Shallis (1983) suggests that sequential time and causality, those central aspects of experience, cease to be meaningful at the sub-atomic level. He writes,

> In turning to smaller and smaller time intervals, a level is reached where causality is no longer distinguishable in nature. This temporal view of the quantum world leads to the insight that causality is itself a temporal phenomenon which only appears at longer time intervals. Causality is a principle that nature displays at the macroscopic level of time. (Shallis, 1983, p. 117)

Whilst quantum physicists too retain time as reversible 't'-coordinate and as quantitative measure, neither the B-relations nor the measure were useful tools for grasping the strange reality they encountered. They found that the Newtonian conceptualisation of the ultimate reality as particles in motion, bouncing off each other like billiard balls, did not describe the temporality of their subject matter. The concept of quantum, explains Capra (1982, pp. 63–85), stands for unit of action which contains energy and time. These quanta are not hard, indestructible bodies, as thought previously, but are better conceptualised as dynamic patterns of activity. They can, in fact, only be observed in terms of their patterned energy interconnectedness. At this level of reality there are no things, signals or causal connections. There are no dancers, only the dance.

It is one of the fundamental insights of quantum theory that *matter cannot be separated from its activities*; and that this activity-matter does not resemble the conventional conceptualisation of inert matter. Quantum physicists describe a 'virtual state' where electrons, for example, seem to try out all their possible futures instantaneously before settling down to exist, and later to disappear again (Zohar, 1983, pp. 127–30). Thus, the social science idea that past, present and future are irrelevant for the description of nature does certainly not apply to the quantum world, and neither does the proposition that the time aspects of nature are quantifiable.

Where physicists take account of interaction, the physical world emerges as asymmetrical and characterised by unidirectional and irreversible time. Towards the end of the last century this entailed a shift of focus from the Newtonian idealisation of particles in motion, and the mathematics of stabilising change, to the principles of change themselves. The fact that energy may be conserved but not re-used for the same work – the meaning of the Second Law of Thermodynamics – forms an integral part of the experience and working knowledge of our everyday lives. Cars rust and use up petrol, homes get untidy, wood burns to ashes, and we need to replenish our body's energy level with food.

All dynamic theories conceptualise energy as being conserved. But in thermodynamics, unlike Newtonian dynamics, energy is conserved but cannot be reversed. This knowledge allows an observer to distinguish processes on a before and after basis. A hot cup of coffee will always cool, unless we do something to prevent it; and shuffling an ordered deck of cards will mess up the order. Thus one knows that the hot state of the cup of coffee precedes the cool state, that the ordered state precedes the shuffled state of the deck of cards, and that the reverse process is so unlikely that we think it impossible. This energy exchange process is just one example of the 'arrow of time' but, as physicists tell us, there are many more natural phenomena that display such unidirectional, irreversible characteristics.[15]

The inorganic world, however, displays irreversible processes not only towards disorder and decay, but also towards new forms and increased order, as the work of Prigogine (1980, and with Stengers, 1984) demonstrates. Prigogine found all processes towards order to entail some form of elementary feedback system, system identity, and self-organisation with an open flow-stability of change-order. Because of these essential characteristics, he argues the case for a conceptualisation of time that can incorporate those aspects. In contradistinction to the Newtonian 't'-coordinate he introduces a time operator 'T', which he wants understood as historical time: as the internal age of a system, which expresses both irreversible directionality, and an essential difference between the past and the future. He established this 'T'-time *as a law of nature*; and with it he changed the very meaning of the nature of a scientific law. As long as laws continue to be conceptualised as 'outside time and space', his dissipative processes cannot be thought about in a meaningful way. Once scientists think of nature as 'T'-time-full, laws themselves come to be understood as developing; and reversibility, far from being the most fundamental aspect of nature, comes to be recognised as a product of the consciousness of the human observer.

The work of this Nobel Prize-winning physical chemist, and its fundamental implications for science, seems thus far to have had no impact on the human sciences. Apart from the German social theorist Luhmann,[16] I am not aware of any social scientists who does not claim the past, present and future aspects of time exclusively for the human realm. Prigogine, however, has established the 'T'-time, and with it both the A- and B-series of time, as a law of nature. It is therefore important that the human sciences take note of this development in the natural sciences.

For an adequate conceptualisation of living systems, many additional aspects of time have to be incorporated. Whilst living organisms entail all the time aspects encountered so far, the latter are no longer sufficient for the understanding, description, and explanation of living phenomena.

Growing, healing, and self-renewal cannot be grasped through structure, or structural relations of time as Capra (1982, ch. 9) and Portmann (1952) have pointed out. They require a conceptual shift from structure to rhythm. They also necessitate an expansion of the time scale so that a multitude of feedback loops may be incorporated in the understanding (Vester, 1979). Living beings, unlike machines, display rhythmicity not only within the organism but also in their interaction with the environment. Whilst rhythmicity is displayed at every level of nature, it is, however, the coming together of a multitude of different rhythms into a whole 'symphony' that distinguishes living organisms from rhythmically oscillating molecules, for example. It is with respect to the symphony of coordinated rhythms that Fraser (1982, ch. 7) writes of the creation of a communal present, of the 'birth of presentness'. Aschoff's (1983) work demonstrates how the existence of the circa-rhythms of the tides, the sun, the moon, and the seasonal rotation of the earth, facilitate anticipatory behaviour. Any organism beating internal time to the rhythms of nature can anticipate, and has the potential for planning the future; it can, as he puts it 'Vorsorge treffen'.[17]

Organisms can thus be conceptualised as clocks in their own right but the nature of the organism clock is different from that of a mechanical one. Neither the nature of its time, nor its knowing is objectively quantifiable; it is fundamentally experiential, relative, and context dependent. It has been established that changes in the metabolic rate, for example, alter the experience of time (Luce, 1977). During hibernation the organism's metabolism is slowed down. Thus, bats may well experience their hibernation as no different from any one night's sleep during the summer, whilst a cold-blooded animal may experience time differently when temperatures vary.

The directional processes encountered in the non-living world have thus to be extended for living beings to encompass growth, ageing, death, and regeneration. Like physical processes, living organisms dissipate energy, but this energy is being transformed and recycled for self-repair, renewal, and creation. The energy dissipated by one organism forms the life-giving energy for another, in a never-ending, never-beginning interconnectedness, where everything affects everything else. The structural relations of the B-series of time do not only prove to be insufficient, but turn out to be meaningless when used as a scientific tool for the study of rhythmic phenomena and ecological interconnectedness. The same applies to the use of Newton's reversible t-coordinate and clock-time. Whilst single durational steps may be measured, living beings can only be grasped in their full rhythmicity and environmental interdependence.

ALL TIME IS SOCIALLY CONSTITUTED – OR IS IT?

I have demonstrated the need to revise some of the conventional social science distinctions between nature and society and their respective times. Having shown quality, the experience of time and past, present and future to be fundamental aspects of nature, it may now be useful to look once more at the idea that 'all time is social time'; a view that is as strongly defended today, as it was at the turn of the century in Durkheim's (1915) *The Elementary Forms of Religious Life*. A summary of the reasoning behind this idea can be expressed in the following way. Firstly, only people regulate and organise their lives by clocks and calendars, only they use, control and allocate time, only they lead an 'in time existence' and create their own histories and futures. Secondly, only humans express meaning symbolically, thus time could not even be thought about without a socially constructed meaning of time.

In response to the first cluster of reasons I would suggest that there are undeniably time aspects which are only found in human social contexts, but that these ought to be distinguished from the more universal principles of time that are to be found throughout nature. Organisational aspects of time, for example, were found at the inorganic, organic, and the human social level, as well as in the design aspects of human artefacts. The organisational principles of time, in terms of sequence, duration, periodicity, rates of change, and synchronisation may be the same for all, but their meaning and their expressive form change with the context. The same applies to an 'in time existence', which is led by simple organisms, plants, animals, and humans alike, because it relates to being bounded by external rhythms. The difference here lies in how the being relates to that rhythm or, in the case of western societies, whether the rhythm has been abstracted from the natural rhythms within which other beings organise their social existences. All beings, it can further be argued, constitute their own past, present, and future. The difference lies in the degree to which they are aware of this fact, and the way they relate to it.

My own research suggests that clock-time, and Newton's time-reversible 't'-coordinate are the only time aspects that can be solely designated as human social constructions. All other aspects, irrespective of whether or not they are conceptualised, are also integral aspects of nature. Natural time is socially constituted in interaction, since the symmetry-breaking process of interaction is one of the sources of time. The idea of time as socially constituted depends thus very much on the meaning we impose on 'the social'; whether we understand it as a prerogative of human social organisation or, following Mead (1959), as a principle of nature.

The idea that time is not separable from the meaning of time, that it always symbolises something that is socially formulated, is a more complex one to untangle. Whilst I do not want to enter into deep philosophical debates of an ontological and epistemological nature, the shortcomings of such a position are sufficiently well known to be briefly mentioned here. If language comes to be equated with reality – even if it does allow for meaning to be context-dependent – then we are no longer left with a basis from which to account for meaning variance, from which to understand non-language-based nature, or from which to translate from one system of meaning to another (Giddens 1976, pp. 17–18). It also leaves no room for understanding novelty and creativity. Putting such an exclusive emphasis on the symbol would further deny meaning and consciousness to the rest of nature. In the light of the most recent scientific understandings of reality this is no longer a tenable position to hold.

At a more substantive level it is possible, as Hägerstrand (1985, pp. 8–12) has pointed out, that social scientists have linked time so intimately to the symbol of time because this is the very form they have been focussing on. But, as this exploration of natural time has shown, time is not exhausted by the symbol of time; it is an integral aspect of nature. I accept that for us to be able to talk and think about time necessitates our putting it into words. If this is all that is being expressed, it is not very much; if it equates reality with the symbol, it goes too far. Thus, whilst it is neither denied that all humans formulate meanings symbolically, nor that this is a fundamentally social process, it is suggested that time is an aspect of nature, and that nature ought to be understood as inclusive of the symbolic universe of human society. Once we recognise ourselves as bearers of all the multiple times of nature, and once we allow for nature to include symbolic expression, then the dualisms of the above understanding are overcome; the gulf between the symbolic knower and nature as an external (unknowable) object can be dispensed with.

From this knowledge we emerge as being activity-matter, causal and non-local communication networks, biological clocks that beat in 'off beat' to the rhythms of our earth: growing and decaying dynamically in interdependence with other systems of change-order. Recognising ourselves as having evolved, and thus being the times of nature, allows for the humanly constituted aspects of time to become one expression among the others. Biologists have dispelled the idea that only humans experience time, or organise their lives by it. Waiting and timing in nature presupposes knowledge of time temporality, irrespective of its being symbolised, conceptualised, reckoned, or measured. Looked at in this way, time may be viewed as having evolved as an aspect of sociality in the universe. The way humans – as a species, or as members of specific societies, groups, and families – symbolise time and relate to it, may then be understood as a specific explicated form of something that is uniquely implicated in all of nature.

CONCLUDING THOUGHTS

Time has been incidental to classical social theory; loosely associated with change, or a mere parameter within which life is enacted. When social scientists focus on time directly, however, the incidental becomes problematic. Suddenly, social scientists find themselves faced with a multitude of times that display a total disregard for established disciplinary boundaries and categories. The rhythms of nature, growing and getting old, contemporary technology, language and social institutions are found to be all mutually implicating and no longer neatly separable into physical, biological and social domains. One response to this dilemma is to turn to classical philosophy for the definition of natural time so that it may once more be safely excluded from social science analysis. If however, we turn instead to the natural sciences for the characterisation of natural time, a fundamentally different picture emerges.

Having followed this path, time as a quantitative measure has emerged as a uniquely human creation, and qualitative time as a fundamental aspect of nature. It seems ironical therefore that this most purely human time should be considered by social scientists as the one most suited for the understanding description, and explanation of the natural world. This is not to suggest that nature cannot be quantified – for whatever human purposes – but that time as measure of motion is an inappropriate tool with which to understand her. Where societies do relate to clock-time as a reality, and where this Newtonian time does form an important part of the social life it does, on the other hand, seem an appropriate tool for social analyses. Since it has been shown that clock-time can never be the only social time, it becomes self-evident that social scientists need to use it in conjunction with other forms of social time.

Nature has further emerged as time-full, and as fundamentally implicating what social scientists deemed specifically human: the past, present, and future. As historical time,

Prigogine's 'T'-time has become a law of nature. No longer timeless, the meaning of a scientific law has changed; it has become recognised as timeful, temporal, and implicating the observer. Intimately tied to understanding nature as timeful is Mead's (1959) recognition of sociality as a principle of nature. Our world, not just human society, is socially constituted. Nature's asymmetry, interactions, transactions, communications, organisms and societies are both expressive of the fact, and source of those irreversible processes through which we understand time as having passed.

This exploration of natural time through the natural sciences has brought to light the social scientists' Newtonian conceptualisation of science, and all that is considered non (human) social time. Their understanding of nature and natural science – borrowed largely from nineteenth-century philosophy – has further been shown to be out of step with that of contemporary natural science.

Without an appropriate understanding of natural time, however, the social time that social scientists seek to understand, study and theorise cannot be grasped. It cannot be grasped either in its distinctness, or in its relatedness to nature. Neither can any of the multiple social aspects of time be understood in their relation to each other. Without an understanding of the full range of aspects of time, it is not possible to establish the nature of the connection between clock-time, chronology, organisational time, historical time and historicity, rhythmicity, time as a resource, the constitution of time in terms of records, possible futures, the difference between the past and the future; a 'time in' and an 'in time' analysis. The symbol which has come to be understood as an independent reality needs to be reconnected to its sources since all of natural time is fundamentally implicated in the multiple social expressions of the time we study.

NOTES

1 A shorter version of this paper was presented to the Third Annual Conference of the Association for the Social Study of Time (ASSET), at Dartington Hall, June 1986. I am

particularly thankful to Jan Adam and George Newell who scrutinised my basic grasp of theoretical biology and physics respectively, and to Paul Atkinson and Alwyn Jones for their very helpful comments on an earlier draft. The research has been conducted with the financial support of an ESRC grant.

2 Since I am presenting a general picture of traditional social science and contemporary natural science views in this paper, the references should be understood to be an exemplary selection only. Thus, the work of Bergmann, 1981; Giddens, 1984; Jaques, 1982; and Lauer, 1981, for example, demonstrates the difficulty of maintaining disciplinary boundaries in the social sciences; Denbigh, 1981; Shallis, 1983; and Zohar, 1983, represent three physicists who have extended their concerns to include issues of consciousness; and the anthologies of Fraser, ed., 1966; Fraser et al., eds, 1972, 1975, 1978, 1981; Fröbe-Kapteyn Hgb, 1952; Landsberg, ed., 1982; Peisl and Mohler Hgb, 1983; Schaltenbrand Hgb, 1963, serve as examples to show that the process of 'joining up' has not yet been successfully completed.

3 I use the term social science in contradistinction to physical and biological science. The majority of individual social scientists I have focussed on, however, have been sociologists.

4 Since I focus on issues that concern a wide range of disciplines, from physics and biology to philosophy and social science, the total composition is likely to make readers feel that their particular field of expertise has received only cursory or simplified treatment, or has even been left out altogether. And so it might, but in order to make visible and meaningful what is conventionally considered outside the domain of social science, I had to select and simplify to suit the purpose of the endeavour. I hope, however, that the result justifies the means.

5 This is demonstrated by the work of Wendorff 1980, essays in the anthologies of Fraser, ed., 1966; Fraser, et al., 1972, 1975, 1978 and 1981, the seminal work of Whorf, 1956; and Evans-Pritchard, 1940; and many other anthropological studies such as Booth Jr, 1975; and Skar, 1981, for example.

6 For general distinctions between the natural and social sciences and their respective subject matters see, for example, Giddens, 1976, Introduction and ch. 4; Lessnoff, 1979; or Ryan, 1970.

7 'Durée' as the continuous emergence of novelty, as that which

constitutes time in ceaseless emergence, has not been successfully incorporated into social theory. On the concepts of 'temps' and 'durée' see Bergson, 1910; and Capek, 1971, especially p. 91. Mead's conceptualisation of the principle of sociality is an exception here. See Mead, 1959, especially chapter 3.

8 These time aspects are specifically focussed on by phenomenologically biased sociologists, see for example Schutz, 1971; Schutz and Luckmann, 1973; and more generally Giddens, 1979, 1981, and 1984.

9 Durkheim first presented the idea in 1915 in response to both Kantian idealist, and pragmatist positions on time. Classical formulations followed by Coser and Coser, 1963; Sorokin and Merton, 1937; and Sorokin, 1964. Bergmann, 1981; Lauer, 1981; Schoeps, 1980; and Zerubavel, 1981 and 1985 represent contemporary approaches of that position, each with slightly different foci and emphases.

10 Elias, 1982; Mead, 1959; and the work of Luhmann, 1978, come to mind here. They are, however, exceptions for different reasons: Elias for having focussed on the shortcomings of traditional philosophy for social science analysis, Mead for having shown the importance of the natural science understanding for social science, and Luhmann for having elaborated the principles of both natural and social systems in their relation.

11 Husserl's distinction between protensions and retentions (Husserl, 1964) and Heidegger's concept of 'Dasein' (Heidegger, 1980) were also used for this purpose, but these were not focussed on in this chapter.

12 It must be appreciated, that where natural scientists attempt a conscious understanding of their own base assumption, they too depend on interpretations, of which there are many. These can be followed up in the anthologies of Fraser, ed., 1966; Fraser et al., 1972, 1975, 1978, 1982; and Landsberg, ed., 1982, for example. Despite this lively debate there is, however, a coherent body of thought emerging whose approach has fundamental implications for a renegotiation of meaning in the social sciences.

13 Bohm, 1983; Capra, 1981 and 1982; Denbigh, 1981; Gardner, 1982; Jones, 1983; Landsberg, ed., 1982; Powers, 1982; Prigogine, 1980, and with Stengers, 1984; Ridley, 1976; Rifkin, 1985; and Shallis, 1983, represent a contemporary trend in theoretical physics to question the

tacitly held knowledge of their discipline's base assumptions.

14 There are proposals for other, more detailed divisions, as shown in the work of Fraser, 1982; and Landsberg, ed., 1982, for example, but these no longer demonstrate with the same clarity the bearing of those physical theories on social ones.

15 See for example the discussion on 'The Arrow of Time' in Powers, 1982, pp. 100–8.

16 This is a view that permeates Luhmann's 1982a, 1982b essays and his work generally, but is most explicitly put in his 1978, 'Temporalization of Complexity'.

17 The German 'Sorge' and 'Vorsorge' are concepts that are particularly timeful. 'Sorge', in which the future orientated care and anxiety are fused to form one word, acquires an additional time aspect with the prefix 'Vor'. As 'Vorsorge' care-anxiety includes the further element of planning. In contradistinction to Heidegger, for whom 'Sorge' is a centrally human characteristic, Aschoff, Portmann, and others show 'Sorge' and 'Vorsorge' to be aspects of all rhythmically organised life.

REFERENCES

Aschoff, J. (1983), 'Die innere Uhr des Menschen', in Peisl, A. and Mohler, A. Hgb., *Die Zeit*, München, Oldenburg Verlag.

Bergmann, W. (1981), *Die Zeitstrukturen sozialer Systeme: Eine systemtheoretische Analyse*, Berlin, Duncker & Humbolt.

Bergson, H. (1910), *Time and Free Will*, London, Swan Sonnenschein.

Bohm, D. (1983), *Wholeness and the Implicate Order*, London ARK Paperbacks.

Booth Jr., N.S. (1975), 'Time and change in African traditional thought', *Journal of Religion in Africa*, vol. 7, pp. 81–91.

Capek, M. (1971), *Bergson and Modern Physics*, Dordrecht, Reidel.

Capra, F. (1981), *The Tao of Physics*, London, Fontana.

Capra, F. (1982), *The Turning Point*, London, Wildwood House.

Coser, L.A. and Coser R.L. (1963), 'Time perspective in social structure', in A.W. Gouldner and H.P. Gouldner (eds), *Modern Sociology*, London, Rupert Hart-Davis, pp. 638–51.

Denbigh, K. (1981), *Three Concepts of Time*, Berlin, Springer-Verlag.

Durkheim, E. (1915), *The Elementary Forms of Religious Life:*

A Study in Religious Sociology, trans. by J.W. Swain, London, George Allen & Unwin.

Elias, N., (1982a), 'Über die Zeit', in *Merkur*, vol. 36, no. 9, September, pp. 841–56.

Elias, N. (1982b), 'Über die Zeit' (part 2) in *Merkur*, vol. 36, no. 10, pp. 998–1016.

Elias, N. (1984), *Über die Zeit*, trans. by H. Fliessbach and M. Schroeter, Frankfurt, A.M., Suhrkamp.

Evans-Pritchard, E.E. (1940), *The Nuer*, Oxford, Clarendon Press.

Fraser, J.T. (ed.) (1966), *The Voices of Time. A Co-operative Survey of Man's Views of Time as expressed by the Sciences and Humanities*, New York, George Braziller.

Fraser, J.T., Haber, F.C. and Mueller, G.H. (eds) (1972), *The Study of Time* I, Berlin, Springer-Verlag.

Fraser, J.T. and Lawrence, N. (eds) (1975), *The Study of Time* II, Berlin, Springer-Verlag.

Fraser, J.T., Lawrence, N. and Park, D. (eds) (1978), *The Study of Time* III, Berlin, Springer-Verlag.

Fraser, J.T., Lawrence, N., and Park, D. (eds) (1981) *The Study of Time*, IV, Berlin, Springer-Verlag.

Fraser, J.T. (1982), *The Genesis and Evolution of Time*, Brighton, Harvester Press.

Fröbe-Kapteyn, O. Hgb. (1952), 'Mensch und Zeit', *ERANOS Jahrbuch 1951*, vol. 20, Zürich, Rhein-Verlag.

Gardner, M. (1982), *The Ambidextrous Universe*, Harmondsworth, Penguin Books.

Giddens, A. (1976), *New Rules of Sociological Method*, London, Hutchinson.

Giddens, A. (1979), *Central Problems in Social Theory*, London, Macmillan.

Giddens, A. (1981), *A Contemporary Critique of Historical Materialism*, London, Macmillan.

Giddens, A. (1984), *The Constitution of Society*, Cambridge, Polity Press.

Hägerstrand, T. (1985), 'Time and culture', in G. Kirsch, P. Nijkamp, and K. Zimmermann (eds), *Time Preference: An Interdisciplinary Theoretical and Empirical Approach*, Berlin, Wissenschafteszentrum.

Heidegger, M. (1980), *Being and Time*, Oxford, Basil Blackwell.

Heirich, M. (1964), 'The use of time in the study of social change', *American Sociological Review*, vol. 29, pp. 386–97.

Husserl, M. (ed.) (1964), *The Phenomenology of Internal Time-*

Consciousness, ed. M. Heidegger, trans. J.S. Churchill with introduction by C.O. Schrag, The Hague, Martinus Nijhoff.
Jaques, E. (1982), *The Form of Time*, London, Heinemann.
Jones, R.S. (1983), *Physics as Metaphor*, London, Abacus.
Landsberg, P.T. (ed.) (1982), *The Enigma of Time*, Bristol, Adam Hilger.
Lauer, R.H. (1981), *Temporal Man. The Meaning and Uses of Social Time*, New York, Praeger.
Lessnoff, M. (1979), *The Structure of Social Science*, London, George Allen & Unwin.
Lewis, J.D. and Weigert, A.J. (1981), 'Structures and meanings of social time', *Social Forces*, vol. 60, no. 2 (December) pp. 433–62.
Luce, G.G. (1977) *Body Time*, St Albans, Paladin.
Luhmann, N. (1978), 'Temporalization of complexity', in R.F. Geyer, and J. van der Zouwen (eds), *Sociocybernetics: An Actor-Orientated Social Systems Approach*, The Hague, Martinus Nijhoff.
Luhmann, N. (1982a), 'The future cannot begin: temporal structures in modern society', in *The Differentiation of Society*, New York, Columbia University Press, pp. 271–89.
Luhmann, N. (1982b), 'World-time and system history', in *The Differentiation of Society*, New York, Columbia University Press, pp. 289–324.
Lüscher, K.K. (1974), 'Time: a much neglected dimension in social theory and research', *Sociological Analysis and Theory*, vol. 4, no. 3, pp. 101–17.
McTaggart, J.M.E. (1927), *The Nature of Existence*, vol. 2, book 5, Cambridge, Cambridge University Press.
Martins, H. (1974), 'Time and theory in sociology', in J. Rex, *Approaches to Sociology*, London, Routledge & Kegan Paul, pp. 246–94.
Mead, G.H. (1959), *The Philosophy of the Present*, ed. A.E. Murphy, La Salle, Illinois, Open Court.
Moore, W.E. (1963), *Man, Time and Society*, New York, John Wiley.
Peisl, A. and Mohler, A. Hgb. (1983), *Die Zeit*, München, Oldenburg Verlag.
Portmann, A. (1952), 'Die Zeit im Leben der Organismen', in Fröbe-Kapteyn, O. Hgb., *ERANOS Jahrbuch* vol. 20, 1951, Zürich, Rhein Verlag.
Powers, J. (1982), *Philosophy and the New Physics*, New York, Methuen.

Prigogine, I. (1980), *From Being to Becoming*, San Francisco, W.H. Freeman.
Prigogine, I. and Stengers, I. (1984), *Order out of Chaos*, London, Heinemann.
Ridley, B.K. (1976), *Time, Space and Things*, Harmondsworth, Penguin Books.
Rifkin, J. and Howard, T. (1985), *Entropy. A New World View*, London, Paladin.
Ryan, A. (1970), *The Philosophy of the Social Sciences*, London, Macmillan.
Schaltenbrand, G. Hgb. (1963), *Zeit in nervenaerztlicher Sicht*, Stuttgart, Ferdinand Enke Verlag.
Schoeps, M. (1980), *Zeit und Gesellschaft*, Stuttgart, Ferdinand Enke Verlag.
Schutz, A. (1971), *Collected Papers* vol. 1, *The Problem of Social Reality*, M. Natanson (ed.), The Hague, Martinus Nijhoff.
Schutz, A. and Luckmann, T. (1973), *The Structures of the Life-World*, trans. R.M. Zaner and H.T. Engelhardt Jr., London, Heinemann.
Shallis, M. (1983), *On Time*, Harmondsworth, Penguin Books.
Skar, S. (1981), 'Andean women and the concept of space/time', in S. Ardener (ed.), *Women and Space*, London, Croom Helm.
Sorokin, P.A. (1964), *Sociocultural Causality, Space and Time*, New York, Russell & Russell.
Sorokin, P.A. and Merton, R.K. (1937), 'Social time: a methodological and functional analysis', *American Journal of Sociology*, vol. 42, no. 5, pp. 615–29.
Vester, F. (1979), 'Time and biology', in F. Greenaway (ed.), *Time and the Sciences*, Paris, UNESCO.
Wendorff, R. (1980), *Zeit und Kultur*, Wiesbaden, Westdeutscher Verlag.
Whorf, B.L. (1956), *Language, Thought and Reality*, Cambridge, Mass., MIT Press.
Zerubavel, E. (1981), *Hidden Rhythms*, Chicago, Chicago University Press.
Zerubavel, E. (1985), *The Seven Day Circle*, New York, Free Press.
Zohar, D. (1983), *Through the Time Barrier*, London, Paladin.

INDEX